SpringerBriefs in Applied Sciences and Technology

SpringerBriefs present concise summaries of cutting-edge research and practical applications across a wide spectrum of fields. Featuring compact volumes of 50–125 pages, the series covers a range of content from professional to academic.

Typical publications can be:

- A timely report of state-of-the art methods
- An introduction to or a manual for the application of mathematical or computer techniques
- A bridge between new research results, as published in journal articles
- A snapshot of a hot or emerging topic
- An in-depth case study
- A presentation of core concepts that students must understand in order to make independent contributions

SpringerBriefs are characterized by fast, global electronic dissemination, standard publishing contracts, standardized manuscript preparation and formatting guidelines, and expedited production schedules.

On the one hand, **SpringerBriefs in Applied Sciences and Technology** are devoted to the publication of fundamentals and applications within the different classical engineering disciplines as well as in interdisciplinary fields that recently emerged between these areas. On the other hand, as the boundary separating fundamental research and applied technology is more and more dissolving, this series is particularly open to trans-disciplinary topics between fundamental science and engineering.

Indexed by EI-Compendex, SCOPUS and Springerlink.

More information about this series at http://www.springer.com/series/8884

Dhananjay Singh · Madhusudan Singh
Zaynidinov Hakimjon

Signal Processing Applications Using Multidimensional Polynomial Splines

Springer

Dhananjay Singh
Department of Electronics Engineering
Hankuk University of Foreign Studies
(Global Campus)
Yongin, Korea (Republic of)

Madhusudan Singh
School of Technology Studies, Endicott
College of International Studies
Woosong University
Daejeon, Korea (Republic of)

Zaynidinov Hakimjon
Head of Department of Information
Technologies
Tashkent University of Information
Technologies
Tashkent, Uzbekistan

ISSN 2191-530X   ISSN 2191-5318   (electronic)
SpringerBriefs in Applied Sciences and Technology
ISBN 978-981-13-2238-9   ISBN 978-981-13-2239-6   (eBook)
https://doi.org/10.1007/978-981-13-2239-6

Library of Congress Control Number: 2018960731

This Springer imprint is published by the registered company Springer Nature Singapore Pte Ltd.
The registered company address is: 152 Beach Road, #21-01/04 Gateway East, Singapore 189721, Singapore

Preface

This book talks about one-dimensional splines and how they offered a faster and more accurate method of approximation and processing of signals. Then it develops that discussion by talking about multidimensional splines and how they provide a much better computational structure as compared to their polynomial counterparts. The book then explains in the following chapter how these newer software like Simulink, MATLAB provide premade functions and a base for developing our own algorithms using them. Finally, the book concludes the study of splines by analysing some of its applications like processing signals to predict scenarios in different fields. In conclusion, splines are a great tool and are highly useful in areas that deal with signals and data processing.

The information contained in this book represents the results of extensive work of the authors at the Saint Petersburg Electrotechnical University, Tashkent Technical University, Tashkent University for Information Technologies, Dongseo University, Hankuk University of Foreign Studies, Endicott College of International Studies and Woosong University, South Korea. The evaluated methods, software and hardware and their practical use are based on a number of scientific research projects.

Yongin, Korea (Republic of)
Daejeon, Korea (Republic of)
Tashkent, Uzbekistan

Dhananjay Singh
Madhusudan Singh
Zaynidinov Hakimjon

About This Book

This book walks one through the idea of splines. We started with the origin of the splines in 1946 formulated by Sheinberg. Firstly, the book talks about one-dimensional splines and how they offered a faster and more accurate method of approximation and processing of signals. Then it develops that discussion by talking about multidimensional splines and how they provide a much better computational structure as compared to their polynomial counterparts. The book then explains in the following chapter how these newer software like Simulink, MATLAB provide premade functions and a base for developing our own algorithms using them. Finally, the book concludes the study of splines by analysing some of its applications like processing signals to predict scenarios in different fields. In conclusion, splines are a great tool and are highly useful in areas that deal with signals and data processing.

Contents

About the Authors

Prof. Dhananjay Singh, SMIEEE is Associate Professor in Electronics Engineering and Director of ReSENSE Laboratory at Hankuk (Korea) University of Foreign Studies (HUFS), Seoul. He is also Chief Technical Officer (CTO) at MtoV Inc., and Technical Director of Vestella, South Korea. He is a ACM distinguished speaker and senior member of ACM and IEEE. He received his B.Tech. degree in computer science and engineering from United College of Engineering and Research (UCER), Allahabad, and M. Tech. degree in wireless communication and computing from IIIT Allahabad, India. He received his Ph.D. degree in ubiquitous IT from Dongseo University (DSU), Busan, South Korea. Before joining the HUFS, he worked as a senior member of engineering staff in the division of future Internet architecture at ETRI, Korea, and PostDoc research for the developing future Internet model at National Institute of Mathematical Sciences (NIMS), Daejeon, South Korea. He is a distinguished speaker of the Forum of Brazil–Korea on Science and Technology, Deployment and Innovation at UNISINOS, Sao Leopoldo, Brazil. He is Author of three books, seven chapters, seven international patents and more than 100 research publications. He has delivered more than 50 invited/keynote talks in IEEE and ACM nad Springer conferences. He is the member of five editorial boards and more than 100 technical program committees (TPC) member in the conferences. His research interests focus on design, analysis and implementation of algorithms and protocols based on network and communication with a specialization in wireless communications, embedded systems, mobile and cloud computing to solve real-world problems based on Internet of things, machine-to-machine communication, smart city technologies and future Internet and 5G Networks.

Madhusudan Singh, Ph.D. SMIEEE is Assistant Professor in Future Technologies of Endicott College of International Studies at Woosong University (WSU), Daejeon, South Korea. Before joining the university, he worked as Senior Engineer in Next-Generation Research and Development Department at Samsung Display, Yongin, from March 2012 to April 2016 and Research Professor in YICT, Yonsei University, South Korea, from June 2016 to February 2018. He received his bachelor's (2003) and master's (2006) degrees in computer application in the State Technical University and master of technology (2008) degree in IT from IIIT Allahabad, India. He received his Ph.D. degree (2012) in the department of ubiquitous IT from Dongseo University, Busan, South Korea. He is a senior member of IEEE and member of many societies such as ACM, IEEE Blockchain, IEEE Transportation. He is serving as Editor/Reviewer/TPC member of several IEEE/ACM conferences and journals. He has published more than 50 technical research papers and more than 10 patents as well as delivered more than 15 invited talks in the fields of blockchain technology and applications, automotive cybersecurity, intelligent vehicles, Internet of things, cloud and mesh networks, mobile display, machine learning and intelligence systems and industrial Internet (Industry 4.0).

Zaynidinov Hakimjon, D. Tech commenced his work in 1984 as Apprentice Researcher in the Department of "Technical Cybernetics" at Tashkent A.R. Beruny Polytechnic Institute. In 1990, he was admitted to the Saint Petersburg State Electrotechnical University (SPETU) Postgraduate Department. In 1993, he finished his Postgraduation with successful candidate's scientific degree defense at the Specialized Scientific Council of SPETU. In 2005, he was conferred on a doctor of science at the Specialized Scientific Council of Tashkent University of Information Technologies (TITU). Over 20 scientific research works of high economic importance have been completed under his supervision or with his participation. He is the author and co-author of 40 textbooks, tutorials and brochures, 300 scientific articles, including over 50 certificates and patents. His scientific works have been published in scientific journals in England, Germany, USA, Slovakia, Japan, South Korea, Malaysia, India, etc. From 2007 to 2009, he worked as Professor at Dongseo University, South Korea. From 2009 to 2011, he worked as Professor and Head of Department of the "Information Technologies" at TITU. Currently he heads the Department of Informational-Communicational Technologies in the management structure of the Academy of State Management at the President of the Republic of Uzbekistan.

List of Figures

List of Tables

Chapter 1
Parabolic Splines based One-Dimensional Polynomial

1.1 The Functional Relationships Based on Parabolic Splines

A number of works have been published to date [1] on the studies of properties of spline functions and their possibilities for technical applications. The broad popularity of spline approximation is explained by its use as a universal instrument for modelling the functions and, as opposed to other mathematical methods at equal informational and hardware costs, they ensure greater accuracy of calculations.

In general, the splines theory is developing in two directions [2]:

- Interpolation splines that satisfy a system of certain limitation conditions and conditions in internal points of fields.
- Smoothing splines, when the issues of optimisation of various types of functionals are studied.

The spline methods are most efficient in discrete tasks of initial data [3]. Let's consider a net Δ at section $[a, b]$:

$$\Delta : \quad a = x_0 < x_1 < \cdots < x_n = b$$

The random degree polynomial spline m of detect d (d—an integer number, $1 \leq d \leq m$) with chains on the net Δ is determined [4] as a function $S_{m,d(x)}$,

(1)

$$S_{m,d(x)} = \sum_{s=0}^{m} d_{i,s}(x - x_i)^s$$

$$x \in [x_i, x_{i+1}], i = 0, 1, \ldots, n - 1 \tag{1.1.1}$$

(2)

$$S_{m,d}(x) \in C^{m-d}[a, b] \tag{1.1.2}$$

© The Author(s), under exclusive license to Springer Nature Singapore Pte Ltd. 2019
D. Singh et al., *Signal Processing Applications Using Multidimensional Polynomial Splines*, SpringerBriefs in Applied Sciences and Technology,
https://doi.org/10.1007/978-981-13-2239-6_1

The derivative from spline of order $(m - d + 1)$ may be broken at $[a, b]$. Therefore, scholars consider different levels of spline smoothness: first order, second order, etc.

Technical applications mostly use low degree splines, particularly the parabolic and cubic splines. The construction process of such splines is considerably simpler than constructing higher-level splines. The equation system matrix that determines spline parameters is three-diagonal with dominating main diagonal, and the efficient methods can be used at solution of the system [5].

1.2   The Theoretical Issues of Splines

The theoretical issues of splines of even- and uneven-level splines are significantly different from each other [6]. For ensuring the existence and uniqueness of an even degree interpolation spline, its nodes must not coincide with the nodes of the interpolation.

(1)   Interpolation sections:

$$\Delta n: \ a = x_0 < x_1 < \cdots < x_n = b, \ n \geq 2;$$

(2)   Spline nodes (possible breakage point's m—of the derivative):

$$\Delta : \ \tilde{x}_0 = a < \tilde{x}_1 < \cdots < \tilde{x}_n < b = \tilde{x}_{n+1}$$

In particular, the function $S_2(x)$ represents an interpolation parabolic spline for function $f(x)$, provided:

(1) $S_2(x) \in P_2$; $x \in (\tilde{x}_i, x_{i+1})$, $i = 0, 1, \ldots, n - 1$;
(2) $S_2(x) \in C^{(1)}[a, b]$;
(3) $S_2(x_i) = f(x_i)$.

The simplest implementation of a spline at $h = 2$, provided the net is even/uniform and the spline nodes $\tilde{x}_i$ are located between the interpolation nodes x_i, $i = 0, 1, 2, \ldots, n - 1$, i.e.

$$\tilde{x}_0 = a$$
$$\tilde{x}_i = \frac{x_{i-1} + x_i}{2}, \quad i = 1, 2, \ldots, n$$
$$\tilde{x}_{n+1} = b$$

If the function $f(x)$ is periodic of $(b - a)$, then it is usually required that spline $S(x)$ must also be periodic $(b - a)$ and have a continuous first derivative at $(-\infty, \infty)$ and that the point $x_0 = a$ is not a node of the spline. Thus, the periodic spline $S(x)$ satisfies the conditions:

$$S_2^{(i)}(a) = S_2^{(i)}(b) \quad (i = 1, 2) \tag{1.2.1}$$

In a general case, the following end conditions are mostly used:

$$S_2'(a) = a_n, \ S_2'(b) = b_n \tag{1.2.2}$$
$$S_2''(a) = A_n, \ S_2''(b) = B_n \tag{1.2.3}$$

where $a_n, b_n, A_n B_n$ are given real numbers.

A specific selection of these numbers depends on the given task. For instance, if the function $f(x)$ has respective derivatives, then it can be placed $a_n = f(a), b_n = f^1(b), A_n = f^{11}(a), Bn = f^{11}(b)$ or replaced by their approximated values of respective derivatives.

If the selection of end conditions is difficult, then it is possible to require that on points $\tilde{x}_1$ и $\tilde{x}_n$, the spline $S_2(x)$ had a continuous second derivative. This is equivalent to the conditions [7]:

$$S_2''(z - o) = S_2(z + o); \ Z = \tilde{X}_i, (i = 1, n) \tag{1.2.4}$$

Assume

$$m_i = S_2''(x_i) \ (i = 0, 1, \dots, n) \tag{1.2.5}$$
$$M_i = S_2''(x_i) \ (i = 0, 1, \dots, n) \tag{1.2.6}$$

If $S_2''(x)$ is a piecewise-constant function, then

$$S_2''(x) = M_i, \ \tilde{x}_i \leq x <= \tilde{x}_{i+1} \quad (i = 0, 1, \dots, n) \tag{1.2.7}$$

Let it be:

$$h_i = x_{i+1} - x_i$$
$$\bar{h}_i = x_{i+1} - \tilde{x}_{i+1}$$

and $f(x_{k-1}, x_k, x_{k+1})$ is the second separate difference of the function $f(x)$ relative to points x_{k-1}, x_k, x_{k+1}.

The formula of recovery of the parabolic spline value [8]:

$$S(x) = f(x_i) + m_i(x - \tilde{x}_i) - C_i(x - x_i)^2 + d_i(x - x_{i+1})^2 \tag{1.2.8}$$

Having required, that $S(x + 1) = f(x + 1)$, $S^1(x + 1) = m_{i+1} \ i=0, 1, 2..., n - 2$, we find the constant values of d_i and c_i.

$$d_i = \frac{f(x_{i+1}) - f(x_i)}{\bar{h}_i(\bar{h}_i - h_i)} - \frac{m_i + m_{i+1}}{2} \cdot \frac{h_i}{\bar{h}_i(\bar{h}_i - h_i)} \tag{1.2.9}$$

$$c_i = \frac{m_{i+1} - m_i}{2h_i} - \frac{f(x_{i+1}) - f(x_i)}{h_i(\bar{h}_i - h_i)} + \frac{m_i + m_{i+1}}{2(\bar{h}_i - h_i)} \qquad (1.2.10)$$

In the periodic case, the equation system looks like the following [9]:

$$\lambda_i' m_{i-1} + (\lambda_i' + \beta_i' + 2)m_i + \beta_i \cdot m_{i+1} = q_i \qquad (1.2.11)$$

where

$$\lambda_{i+1}' = \frac{h_{i+1}(h_i - \bar{h}_i)}{\bar{h}_i(h_i + h_{i+1})}; \beta_{i+1}' = \frac{\bar{h}_{i+1} \cdot h_i}{(h_{i+1} - \bar{h}_{i+1})(h_i + h_{i+1})} \qquad (1.2.12)$$

$$\lambda_{i+1} = \frac{h_{i+1}}{h_i + h_{i+1}}; \beta_{i+1} = \frac{h_i}{h_i + h_{i+1}}; \qquad (1.2.13)$$

$$q_i = 2\lambda_i \frac{f(x_i) - f(x_{i-1})}{\bar{h}_{i-1}} + 2\beta_i \frac{f(x_{i+1}) - f(x_i)}{h_i - \bar{h}_i} \qquad (1.2.14)$$

where it should be assumed, $m_0 = m_n$, $m_{n+1} = m_0$, $h_n = h_0$, $\bar{h}_n = \bar{h}_0$.

Thus, construction of a parabolic spline as per (1.2.10) consists of finding the values of m_i by solution of equation systems, the general appearance of which is:

$$Az = q \qquad (1.2.15)$$

Their matrices are banded in all cases with diagonal prevalence; i.e. if a_{ij} is an element of the ith line and jth column, then

$$r_i = |a_{ii}| - \sum_{j \neq l} |a_{ij}| > 0$$

for all i.

We shall consider an equation system with a three-diagonal matrix [10].

$$\begin{vmatrix} a_1 & b_1 & 0 & \ldots & & 0 & C_1 \\ C_2 & a_2 & b_2 & \ldots & & 0 & 0 \\ \ldots & \ldots & \ldots & \ldots & \ldots & & \\ 0 & 0 & 0 & & C_{N-1} & a_{N-1} & b_{N-1} \\ b_N & 0 & 0 & & 0 & C_N & a_N \end{vmatrix} \qquad (1.2.16)$$

The system solution algorithms (1.2.15) using the sweep method are given in Chap. 3.

Once the values of m_i are found, taking into account (1.2.9) and (1.2.10), the calculations can use the spline representations in (1.2.8) form. In this case, it is required that the values of $\{x_i\}$, $\{f(x_i)\}$ $\{\tilde{x}_i\}$ $\{m_i\}$ must be memorised. From (1.2.10), there is $S''(X_i) = 2C_i$; therefore, at $[\tilde{x}_i, \tilde{x}_{i+1}]$ $i=0, 1\ldots, n$ the representation of can be used.

$$S(x) = f(x_i) + m_i(x - x_i) + c_i(x - x_i)^2 \qquad (1.2.17)$$

Assessment of interpolation accuracy using the parabolic spline of the triple-differentiated function of $f(x)$ is determined by inequality:

$$\sum \leq \frac{\sqrt{3}}{216} \max \left| f'''(x) \right| h_i^3 \qquad (1.2.18)$$

Any spline of sufficient smoothness can be represented by basic splines. Particularly, at $d=1$ for expansion, so-called normalised basic splines of level m (B-splines) are used. These are local (finite), piecewise-polynomials functions and satisfy the following conditions [11–13]:

(1) $B_m(x) \equiv 0$ at $x \notin (X_i, X_{i+m+1})$;
(2) $B_m(x) > 0$ at $x \in (X_i, X_{i+m+1})$;
(3) $\int\limits_a^b B_m(r)dr = \int\limits_{X_i}^{X_{i+m+1}} B_m(r)dr = 1.$

Their Fourier transformation as a finite function is determined by the formula [14]:

$$F_m(w) = \left(\frac{Sin(w/2)}{w/2} \right)^{m+1}$$

They can also be determined by the results of convolution/digest operation of B-splines of lower orders:

$$B_m(x) = B_{m-i}(x) \cdot B_0(x) = \int\limits_{-\infty}^{\infty} B_{m-1}(x)B_0(x - r)dr \qquad (1.2.19)$$

where $B_0(r)$ is rectangular impulses (rect-functions).

For ensuring approximation at the whole interval $[a, b]$, B-splines must be given at a wider field of inclusion of $2m$ additional nodes $i=-m, m+1,\ldots, n+m$, while all nodes can be located unevenly.

For calculation of normalised B-splines of a random level m, the recursive formulae can be used [15]:

$$B_{m,i} = \begin{cases} \frac{x-x_i}{x_{i+m}-x_i} \cdot B_i^{m-1}(x), \\ \frac{x_{i+m+1}-x_i}{x_{i+m+1}-x_{i+1}} \cdot B_{i+1}^{m-1}(x), \end{cases}$$

$$x \in \left[x_i - \frac{m+1}{2}, x_i + \frac{m-1}{2} \right]$$

$$x \in \left[x_i - \frac{m-1}{2}, x_{i+1} \frac{m+1}{2} \right]$$

The simplest analytical values for B-splines are constructed for cases of uneven representation of nets. We shall view these representations for basic elements of some levels.

(1) For the first-level B-splines

$$B_{0,1}(x) = \begin{cases} x - 1, & \text{at } x \in [-1, 0] \\ 1 - x, & \text{at } x \in [-1, 0] \\ 0 \end{cases} \tag{1.2.20}$$

(2) For the second-level B-splines

$$B_{0,2}(x) = \begin{cases} 0, & x \geq 3/2 \\ \frac{1}{2}\left(\frac{3}{2} - x\right)^2 & 1/2 \leq x < 3/2 \\ \frac{3}{4} - x & 0 \leq x < \frac{1}{2} \\ B_{0,2}^0(-x) & x < 0 \end{cases} \tag{1.2.21}$$

The second-level B-spline nodes are located on points $i - 1/2$, $(i = -2, -1, 0, \ldots, n)$. Selection of nodes in the middle between interpolation nodes is caused by the requirements of the uniqueness of approximation.

The methodological inaccuracy of approximation by parabolic splines is determined by inequality:

$$\varepsilon \leq \frac{\sqrt{3}}{216} \max \left| f'''(x) \right| \cdot h^3$$

1.3 Evaluation of Parabolic B-Splines of Multiple Levels

Figure 1.1 shows the graphics of the zero, first- and second-level B-splines for uneven net cases. The zero-level B-spline graphics coincides with the widely known rect-functions.

One of the important properties of B-splines is continuity of their several derivatives. For instance, the first and second derivatives from the parabolic basic B-spline:

(1) at interval $-1.5 \leq x < -0.5$

$$B_{0,2}(x) = \frac{1}{2}(1.5 - x)^2 = 0.5\left(2.25 + 3x + x^2\right);$$
$$B_{0,2}'(x) = 0.5(2x + 3) = 1.5 + x;$$
$$B_{0,2}''(x) = 1;$$

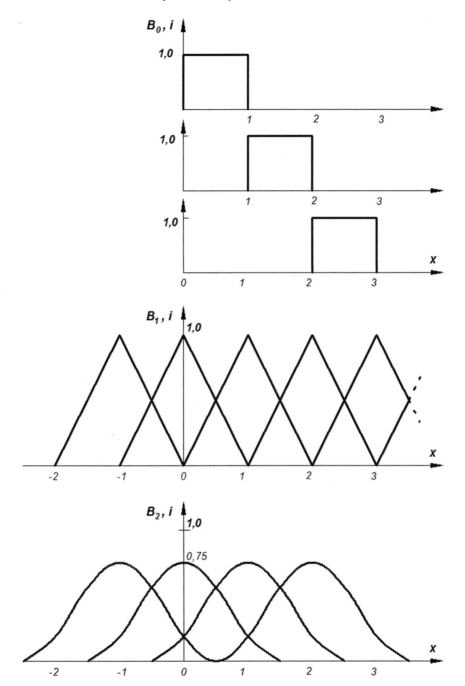

Fig. 1.1 Parabolic B-splines of zero, first and second levels

(2) at interval $-0.5 \le x < 0.5$

$$B_{0.2}(x) = \frac{3}{4} - x^2;$$
$$B'_{0.2}(x) = -2x;$$
$$B''_{0.2}(x) = -2;$$

(3) at interval $0.5 \le x < 1.5$

$$B_{0.2}(x) = \frac{1}{2}(1.5 - x)^2;$$
$$B'_{0.2}(x) = -1.5 + x;$$
$$B''_{0.2}(x) = 1;$$

Figure 1.2 shows the graphics of basic B-splines of the second level and their first and second derivatives. Thus, an analysis of the methods of functional relationship approximation by parabolic splines allows making the following conclusions:

1. Interpolation with parabolic splines requires the solution of algebraic equation systems of type (1.2.17), but their matrices are incomplete (but three-diagonal instead) for the solution of such systems, there are efficient algorithms (sweep algorithms), as opposed to interpolation using classical polynomials.
2. An important property of the parabolic splines is continuity of their derivatives of the first and second order. This feature may be used for evaluation of a hardware-oriented algorithm for calculation of coefficients in piecewise-quadratic bases, which allows achieving high accuracy of approximation of signals and functions.
3. There are two types of representation of splines: polynomial type (Formula 1.2.19) and basic type (Formula 1.2.21). The disadvantage of polynomial splines is that they cannot be well parallelised, and the algorithms obtained on their basis are not hardware-oriented and, consequently, the polynomial splines are not convenient from the point of view of hardware implementation. The advantage of the basic splines is both analytical and tabular representation of given functional relationships by parabolic splines of type (1.2.11) that allows dividing the initial relationship into two parts: the first being the basic function, local and independent from the initial relationship, while the second represents coefficients that depend upon the initial relationship. Formula (1.2.21) is convenient for hardware implementation and represents serious opportunities for parallelising calculations.

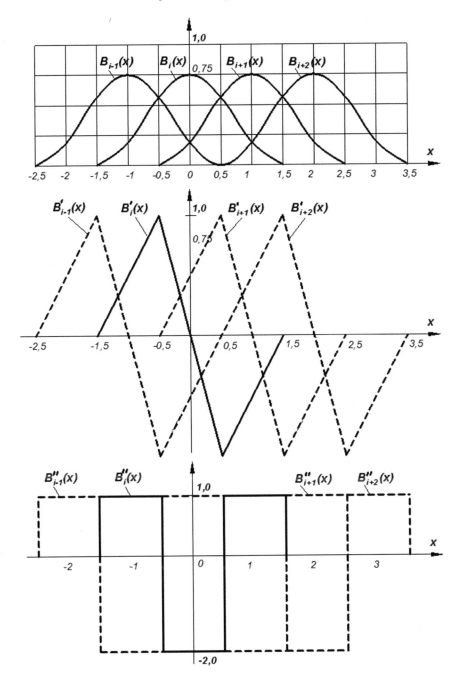

Fig. 1.2 Parabolic B-splines and their derivatives of the first and second levels

1.4 Summary

An analysis of spline methods of approximation identified that most of the practically used elementary functions could be successfully approximated by basic splines. The mathematical installation of approximation with basic splines allows representing functional relationships as sums of pair derivatives of constant coefficients and the basic function values. This gives a base for significant parallelising of calculation both analytical and tabular given functional relationships. The local property of basic splines determines a limited number ($N + 1$, where N is the spline level) of summands in approximating sums and a minimal scope of basic function values' tables. Broad automation based on computer technologies is a specific feature of modern stage scientific research work. This is associated with large amounts of data to be processed. The new methods and algorithms gain more importance for they ensure timely and efficient data transformation. The contemporary signal and image processing methods are mostly dependent on the development of algorithmic and structural means, and elemental and architectural aspects of computing facilities.

References

1. M. Antonelli, C. Beccari, G. Casciola, A general framework for the construction of piecewise-polynomial local interpolants of minimum degree. Adv. Comput. Math. (2013). https://doi.org/10.1007/s10444-013-9335-y
2. C.V. Beccari, G. Casciola, S. Morigi, On multi-degree splines. Comput. Aided Geom. Des. 58 (2017). https://doi.org/10.1016/j.cagd.2017.10.003
3. F. Johnson, M.F. Hutchinson, C. The, C. Beesley, J. Green, Topographic relationships for design rainfalls over Australia. J. Hydrol. **533**, 439–451 (2016). ISSN 0022-1694, https://doi.org/10.1016/j.jhydrol.2015.12.035
4. S. Reboul, M. Benjelloun, Fusion of Piecewise Stationary Process.(2002). https://www.researchgate.net/profile/S_Reboul/publication/254461483_Fusion_of_piecewise_stationary_process/links/540eac620cf2f2b29a3a923d.pdf?origin=publication_detail
5. D. Singh, H. Zaynidinov, H.J. Lee, Piecewise-quadratic Harmut basis funictions and their application to problem in digital signal processing. Int. J. Commun. Syst. **23**, 751–762 (2010). (www.interscience.wiley.com). https://doi.org/10.1002/dac.1093
6. W. Wang, Special quadratic quadrilateral finite elements for local refinement with irregular nodes. Comput. Methods Appl. Mech. Eng. **182**(1–2), 109–134 (2000). ISSN 0045-7825. https://doi.org/10.1016/S0045-7825(99)00088-2
7. Y.V. Zakharov, T. Tozer, Local spline approximation of time-varying channel model. Electron. Lett. **37**, 1408–1409 (2001). https://doi.org/10.1049/el:20010942
8. B.I. Kvasov, Parabolic B-splines in interpolation problems. USSR Comput. Math. Math. Phys. **23**(2), 13–19 (1983). ISSN 0041-5553. https://doi.org/10.1016/S0041-5553(83)80041-X
9. S. Gao, Z. Zhang, C. Cao, Differentiation and numerical integral of the cubic spline interpolation. J. Comput. **6**(10), 2037–2044 (2011)
10. T. Blu, M. Unser, Wavelets, fractals, and radial basis functions. IEEE Trans. Signal Process. **50**(3), 543–553 (2002)

11. Y. Lipman, D. Levin, D. Cohen-Or, Green Coordinates. In ACM SIGGRAPH 2008 papers (SIGGRAPH '08). ACM, New York, NY, USA, Article 78, 10 pp (2008). https://doi.org/10.1145/1399504.1360677
12. B. Mohandes, Y.L. Abdelmagid, I. Boiko, Development of PSS tuning rules using multi-objective optimization. Int. J. Electr. Power Energy Syst. **100**, 449–462 (2018). ISSN 0142-0615. https://doi.org/10.1016/j.ijepes.2018.01.041
13. M. Walz, T. Zebrowski, J. Küchenmeister, K. Busch, B-spline modal method: a polynomial approach compared to the Fourier modal method. Opt. Express **21**(12), 14683–14697 (2013)
14. O. Hidayov, D. Singh, B. G.B. Gwak, S-Y. Young, A Simulink-model of specialized processor on the piecewise-polynomial bases. International Conference on Advanced Communication Technology, ICACT (2011)
15. G. Canan Hazar, M. Ali Sarıgöl, Acta Appl. Math. **154**, 153 (2018). https://doi.org/10.1007/s10440-017-0138-x

Chapter 2
B-Spline Approximation for Polynomial Splines

2.1 Specialised Computing Structure for Running B-Spline Approximation

The traditional tabular-polynomial methods for the approximation of complex functional relationships have a limitation on accuracy, smoothness, and they do not allow parallelisation of the computing structures for the implementation of approximation using classical polynomials on Horner's algorithm [1–3]. For instance, a second-level polynomial on Horner's algorithm can be represented in the following way:

$$P_2(x) = a_2 x^2 + a_1 x + a_0 = ((a_2 x + a_1)x + a_0) \tag{2.1.1}$$

The structure, shown in Fig. 2.1, was proposed for the implementation of the second-level polynomials. The use of parabolic basic B-splines for the approximation of functions allows enhancing the smoothness and locality properties [4, 5]. According to Formula (2.1.1), the value of the function under interpolation on a random point of a given interval is determined by the values $m + 1$ of summands (where m is the B-spline level)-pair derivatives of basic functions to constant coefficients.

The local formulae are convenient among the coefficient determination methods in data processing with B-splines for they allow avoiding solution of equation systems [6]. At calculation of b coefficients using the three-point formula, only three values f_{i-1}, f_i, f_{i+1} of the function are used. This facilitates the initiation of processing without waiting for the arrival of all data [7], i.e. a combination of data entry and processing operations.

© The Author(s), under exclusive license to Springer Nature Singapore Pte Ltd. 2019   13
D. Singh et al., *Signal Processing Applications Using Multidimensional Polynomial Splines*, SpringerBriefs in Applied Sciences and Technology, https://doi.org/10.1007/978-981-13-2239-6_2

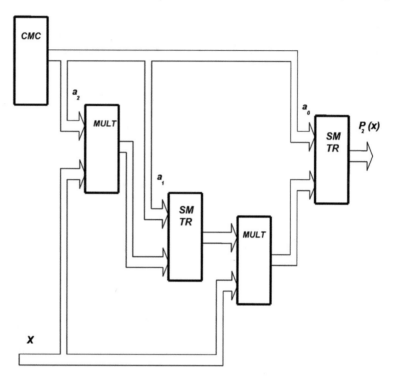

Fig. 2.1 A tabular-algorithmic computing structure for the implementation of approximation of functions using the second-level classical polynomials

The local properties of the basis are demonstrated in this way, and the structure may be constructed in accordance with tabular-algorithmic methods. The use of parabolic B-splines would require three basic summands [8]. We shall assume that the values of this argument are fit to the range [0, 1], and the function values are calculated using the formula:

$$f(x) \cong S_2(x) = b_{-1} \cdot B_{-1}(x) + b_0 \cdot B_0(x) + b_1 B_1(x) \tag{2.1.2}$$

The rest of the basic B-splines at this subinterval are equal to zero, and therefore, they do not participate in the construction of the amount [9]. Formula (2.1.1) indicates the necessity to add (sum) the pair derivatives in groups of three and periodically update the sums that are accumulated in the adder.

The structural diagram (Fig. 2.1) consists of an argument RgAr registry, a coefficient memorising unit (MU) that consists of three registers (Buf, Buf2 and Buf3), a basis functions memory installation (BFMI), three adders and two accumulating adders. For all the basic splines are similar in form, then, using the PMI as BFMI, the task may be limited by memorising the values table of one basic spline only [10]. Figure (2.2) represents a basic parabolic spline, which is determined as zero

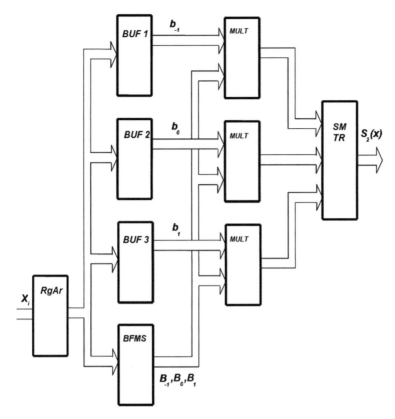

Fig. 2.2 A tabular-algorithmic computing structure for the implementation of approximation of functions using the second-level basic splines

at three areas of piecewise segmentation. For a set of various values of B-splines for the purposes of summing on formula (2.1.2), some three subsections of MU are required, and those subsections are part of the B-spline curve that is given at on area, in each one of such MI. Let us indicate the selected values of argument $X = Xnp$ and assume that it belongs to the area [0, 1]. Three ordinates on curves at different basic splines B_{-1}, B_0, B_1 correspond to this argument. Should the value codes of one only basic spline be entered in the PMI, the values of other basic splines can be determined using two charges of a binary code from the large charges, because the arguments at other three areas are displaced in relation to the Xnp at a distance along the obscys axis, equal to integer numbers (Fig. 2.3). Thus, the imposition of an address group of three values of basic splines requires only two additional binary codes [11], which help to determine the one area out of three of the carrier of the main basic spline, that hosts the values B_{-1}, B_0, B_1 (Figs. 2.4 and 2.5).

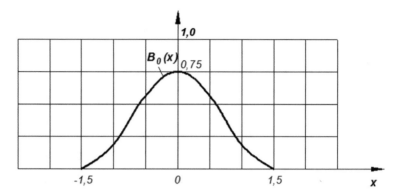

Fig. 2.3 A parabolic basic spline

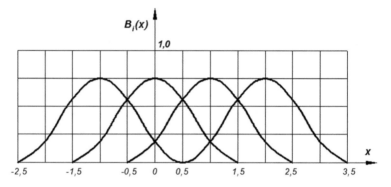

Fig. 2.4 Families of parabolic basic splines [10]

The same subword of the address that indicates an area of $h = 1$ length can be used for the selection of basic spline values in each of three subsections of PMI, which can be given by the following addresses:

$$1\ 1\ \text{for } x \in [-1.5, -0.5]$$

$$0\ 0\ \text{for } x \in [-0.5, 0.5]$$

$$0\ 1\ \text{for } x \in [0.5, 1.5]$$

A matrix diagram of the functioning of the structure for the implementation of approximation of functions based on classical polynomials is shown in Fig. 2.6. In a comparison of the computing structures that implement approximations with classical polynomials with the basic splines [12], it is easy to see that the second

Data entry cycles	Processing cycles				
	1	2	3	4	5
1	Bb a_2 Bb x				
2		mult $a_2 x$ Bb a_1			
3			Add $a_2 x + a_1$ Bb x		
4				Mult $(a_2 x + a_1)x$ Bb a_0	Add $(a_2 x + a_1)x + a_0$

Fig. 2.5 A matrix diagram of the functioning of tabular-algorithmic computing structure on the basis of the second-level classical polynomials [11]

structure is faster than the first. Approximation of a function value on one point takes, in the case of a second-level polynomial $t = 1$ mcs, in the case of using local spline approximation with basic parabolic splines, the time consumed would be $t = 600$ ns only (at constant command cycle of 200 ns).

Data entry cycles	Processing cycles		
	1	**2**	**3**
1	Bв b_{-1}, b_0, b_1 Bв B_{-1}, B_0, B_1		
2		Multiply $b_{-1} B_{-1}$ Multiply $b_0 B_0$ Multiply $b_1 B_1$	
3			Add $b_{-1} B_{-1} + b_0 B_0 + b_1 B_1$

Fig. 2.6 A matrix diagram of functioning of tabular-algorithmic computing structure on the basis of the second-level splines

2.2 Summary

In addition, higher requirements are posed to the specialized computer systems with regard to functioning capabilities in real time (RT), use of computing multisequencing and pipelining principles, and in bandwidth expansion of signals under processing and recovery. Thus, lots of research institutes and scientists have allocated their time and resources to work on this topic.

Signals are reported from gauges installed at various places as data about the status and measurement of temperature, radiation, electromagnetics, gravity, heat and other physical fields is often multidimensional and complex. The requirements for the production rate of computer systems in these fields could be met owing to both evaluation of new methods and algorithms for digital processing of signals (DPS) and signal multiprocessor parallel-pipelining calculation facilities.

The spline functions and generalised spectral methods are widely used for the analysis and recovery of signals. Functions that are glued from various pieces of polynomials on a fixed system are called splines. The obtained smooth homogeneous structure piecewise-polynomial functions (compilation from polynomials of the same degree) are called spline functions or simply splines. They are a developing field of the function approximation and digital analysis theory. The broken spline function is the simplest and the most historical example of splines.

References

1. D. Amadori, L. Gosse, *Error estimates for well-balanced schemes on simple balance laws: one-dimensional position-dependent models*. BCAM Springer Briefs in Mathematics (2015)
2. A. Jagannathan, R. Orbach, O. Entin-Wohlman, *Chapter 4 Thermal Conduction due to Hopping Processes in Amorphous Solids,* ed. by M. Pollak, B. Shklovskii, Modern Problems in Condensed Matter Sciences, vol 28 (Elsevier, 1991), pp. 125–141. ISSN 0167-7837, ISBN 9780444880376. https://doi.org/10.1016/B978-0-444-88037-6.50010-5
3. B. Hofner, A. Mayr, N. Robinzonov et al., Comput Statistics **29**(1–2), 3–35 (2014). https://doi.org/10.1007/s00180-012-0382-5
4. X. Jia, P. Ziegenhein, S.B. Jiang, GPU-based high-performance computing for radiation therapy. Phys. Med. Biol. **59**(4), R151–R182 (2014). https://doi.org/10.1088/0031-9155/59/4/r151
5. N. Brisebarre, S. Chevillard, M. Ercegovac, J.-M. Muller, S. Torres, *An Efficient Method for Evaluating Polynomial and Rational Function Approximations*, (2008) pp. 233–238. https://doi.org/10.1109/asap.2008.4580185
6. X. Liu, S. Guillas, M.J. Lai, Efficient spatial modeling using the SPDE approach with bivariate splines. J. Comput. Graphical Stat. **25**(4), 1176–1194 (2016). https://doi.org/10.1080/10618600.2015.1081597
7. D. Inman, R. Elmore, B. Bush, A case study to examine the imputation of missing data to improve clustering analysis of building electrical demand. Building Serv. Eng. Res. Technol. **36**(5), 628–637 (2015). https://doi.org/10.1177/0143624415573215
8. C. Beccari, G. Casciola, M.-L. Mazure, *Design or not design?* (Numerical Algorithms, A numerical characterisation for piecewise Chebyshevian splines, 2018). https://doi.org/10.1007/s11075-018-0533-z
9. A. Grigorenko, S. Yaremchenko, Investigation of static and dynamic behavior of anisotropic inhomogeneous shallow shells by Spline approximation method. J. Civil Eng. Manage. **15**(1), 87–93 (2009). https://doi.org/10.3846/1392-3730.2009.15.87-93
10. B. Despres, Ch. Buet, The structure of well-balanced schemes for Friedrichs systems with linear relaxation. Appl. Math. Comput. **272**, 440–459 (2016)
11. C.G. Zhu, X.Y. Zhao, Self-intersections of rational Bézier curves. Graph. Models **76**(5), 312–320 (2014). https://doi.org/10.1016/j.gmod.2014.04.001
12. Laurent Gosse, $\mathscr{L}$-splines and viscosity limits for well-balanced schemes acting on linear parabolic equations. Acta Applicandae Mathematicae, Springer **153**(1), 101–124 (2018). https://doi.org/10.1007/s10440-017-0122-5

Chapter 3
One-Dimensional Polynomial Splines for Cubic Splines

3.1 Parallel Computing Structure Based on Cubic Basic Splines

Similarly, to the case of the second-level polynomial, the classical third-level polynomial on Horner's method can be represented in the following way:

$$P_3(x) = a_3 x^3 + a_2 x^2 + a_1 x + a_0 = (((a_3 x + a_2)x + a_1)x + a_0) \qquad (3.1.1)$$

There is a structure for implementation of cubic basic splines polynomial. It is illustrated in Fig. 3.1. However, it is not maximally parallel. Basic splines should be used for improving the performance owing to the parallelising implementation of multiplication operations [1].

According to the formula (2.1.1), the value of function under interpolation at a random point in a given interval is determined by the values of the only $m + 1$ of

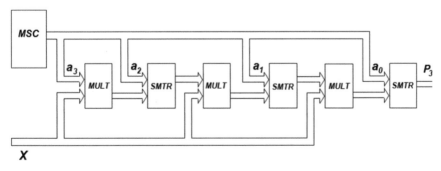

Fig. 3.1 Tabular-algorithmic computing structure for implementation of approximation of functions with the third-level classical polynomials

© The Author(s), under exclusive license to Springer Nature Singapore Pte Ltd. 2019
D. Singh et al., *Signal Processing Applications Using Multidimensional Polynomial Splines*, SpringerBriefs in Applied Sciences and Technology,
https://doi.org/10.1007/978-981-13-2239-6_3

summand-pair derivatives of basic splines by constant coefficients. For instance, the cubic B-splines require four basic summands.

The function value is calculated using the formula

$$f(x) \cong S_3(x) = b_{-1}B_{-1}(x) + b_0 B_0(x) + b_1 B_1(x) + b_2 B_2(x) \Pi x \in [0, \ 1] \quad (3.1.2)$$

The rest of the basic splines at this subinterval are equal to zero and, consequently, do not participate in the formation of the sum.

Different methods exist for calculation of b-coefficients: interpolation and "points" formulae, smoothing splines and the smallest quadrates method [2–4]. However, the "points" formulae should be used for the systems that function in real time. The specific feature of these methods is in the independence of the approximating spline value in the given area from the experimental function values that are remote from the given area [5]. Below are the ready-made "points" formulae for the cubic splines:

1. *Three-points formula*, in this case, the $r - 1$, r and $r + 1$th values of the function participate in determination of the coefficient

$$b_r = \frac{1}{6}(-f_{r-1} + 8f_r - f_{r+1}) \tag{3.1.3}$$

2. *Five-points formula*, in this case, the $r - 2$, $r - 1$, r, $r + 1$ and $r + 2$th values of the function participate in determination of the coefficient

$$b_r = \frac{1}{36}(f_{r-2} - 10f_{r-1} + 54f_r - 10f_{r+1} + f_{r+2}) \tag{3.1.4}$$

Here f_r reduced a form of record—$f_r(x)$.

The methodological inaccuracy of function interpolation $f(x)$ using cubic basic splines is determined by inequality:

$$\varepsilon \leq \frac{5}{384} h^4 \max |f^{IV}(x)| \tag{3.1.5}$$

For function $f(x) = \ln(1 + x)$, we shall produce

$$\varepsilon \leq \frac{5}{384 \cdot 1.0 \cdot 32^4} = 0.12 \cdot 10^{-7} \tag{3.1.6}$$

As a comparison, we shall view the inaccuracy value of interpolation with Newton's classical cubic splines:

$$\varepsilon \leq \frac{1}{24} h^4 \max |f^{IV}(x)| = \frac{1}{24 \cdot 32^4 \cdot 1.0} = 0.4 \cdot 10^{-7} \tag{3.1.7}$$

On (3.1.6), we see that the inaccuracy is over three times higher than the value, obtained in (3.1.7).

The structure of a tabular-algorithmic compuyting structure for implementation of approximation of function with cubic basic splines consists of an argument register (ARg), a coefficient memorising unit (CMU), a basic function memorising unit (BFMU) and an outlet summation-accumulator of accumulating summations [6]. For all the basic splines are similar in form, then, using the CMU block as BFMU, it can be limited by memorising the values table of only one basic spline [7, 8]. For a selection of four different values of B-splines for the purposes of summing on formula (3.1.1), four MU subsections will be required, each containing a part of the B-spline curve, given in this area.

Similarly, with the parabolic splines case, application of an address of four values of basic splines requires only two additional binary bits, which help determine the one out of the four areas of the carrier of the main basic spline, which hosts the values B_{-1}, B_0, B_1 и B_2. The same subword of the address that indicates an area of $h = 1$ length, can be used for selection of basic spline values in each of three subsections of PMU, which can be given the following addresses:

$$11 \text{ for } B_{-1} \ x \in [-2, \ -1];$$
$$00 \text{ for } B_0 \quad x \in [-1, \ 0];$$
$$01 \text{ for } B_1 \quad x \in [0, \ 1];$$
$$10 \text{ for } B_2 \quad x \in [1, \ 2].$$

For confirmation of justification of the conclusion on applicability of hardware costs as a super large-scale integration circuit memory and their slight increase at rise of the number of simultaneously performed functions at one installation, we shall divide the whole task interval of the random, multiple differentiated function $f(x)$ into areas at n $- 2^0$, where p is an integer number. Then, if the binary code of the argument contains l bits, [9] one approximation area would have 2^{i-1} values of elements of the basic spline table, while the total scope of the table is determined using the formula:

$$Q = 4 \cdot 2^{l-1-p} = 2^{l-p+1}$$

Due to the evenness of the basic spline, only half of the table can be recorded in the memory, for instance for the range of argument [0, 2].

If $n = 32$, while $l = 16$ of binary charge, then $Q = 2^{16} - 2^5 = 2048$ words, which at capabilities of modern super large-scale integration circuits, can be implemented using one or two microchips. The storage capacity coefficients at transformation to one function would be equal to $n + 2m + 1$ words.

Reconstruction of a functional relationship to a new shape is done by simple transfer of coefficients to a new storage capacity area. Upon recoding of basic function in PMU, generally, the algorithm of transformation is linear [8]. A parallel structure can be expanded in case of necessity in an increase of the basic spline level. However,

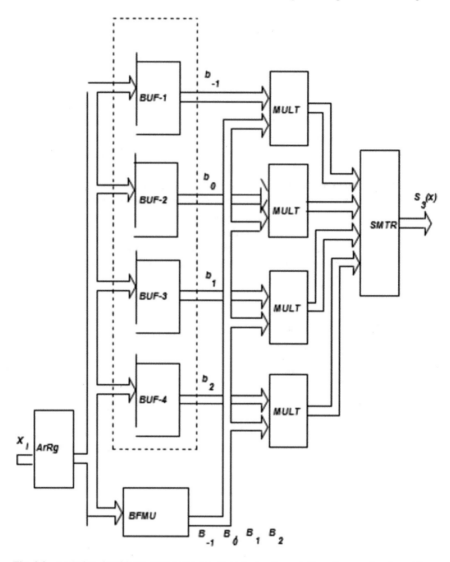

Fig. 3.2 A tabular-algorithmic computing structure for implementation of approximation of function with cubic basic splines

this leads to complication of data output from memory and increase of the number of multipliers.

Thus, the main advantage of the structure is its high performance, actually top speed for tabular-algorithmic methods (Fig. 3.2).

3.2 Summary

Different approximation methods using splines can be used for digital processing and recovery of signals of geophysical information and bench test works. The use of methods depends on the problem. For systems that function in real time, a method is proposed for calculating coefficients using the "points" formulae, which helps avoid solving equation systems. It was proved that maximum rounding inaccuracy at approximation with a second-level spline is approximately four times less than its classical second-level polynomial counterpart. It was also proved that the methodological inaccuracy with classical cubic polynomials is over three times higher than it is with cubic basic splines.

Owing to the universality of readout (observation) processing algorithms, good differential and extreme properties, good convergence rate of approximation assessments, simplicity of calculations of forms and parameters, insignificant influence of rounding errors, as a class of piecewise functions, splines, are more widely used in creation of hardware and software for analysis and recovery of one-dimensional and multidimensional signals, thus expanding the traditional approach ranges. The use of obtained theoretical results, implementation of developed software and hardware, and using such, solution of analytical tasks of processing and recovery of signals and approximation of functions allow modification of existing and evaluation of new computing structures with more potential, enhance their technical specifications and improve the quality of the results of data processing and recovery.

References

1. A.G. Akritas, S.D. Danielopoulos, On the complexity of algorithms for the translation of polynomials. Computing **24**(1), 51–60 (1980). https://doi.org/10.1007/BF02242791. (Springer)
2. O. Sobrie, N. Gillis, V. Mousseau, M. Pirlot, UTA-poly and UTA-splines: additive value functions with polynomial marginals. Eur. J. Oper. Res. **264**(2), 405–418, ISSN 0377-2217, (2018). https://doi.org/10.1016/j.ejor.2017.03.021
3. J. Goh, A.A. Majid, A.I.M. Ismail, Numerical method using cubic B-spline for the heat and wave equation. Comput. Math. Appl. **62**(12) (December 2011), 4492–4498 (2011). http://dx.doi.org/10.1016/j.camwa.2011.10.028
4. S. Jana, S. Ray, F. Durst, A numerical method to compute solidification and melting processes. Appl. Math. Model. **31**(1), 93–119, ISSN 0307-904X, (2007). https://doi.org/10.1016/j.apm.2005.08.012
5. X. Jia, P. Ziegenhein, S.B. Jiang, GPU-based high-performance computing for radiation therapy. Phys. Med. Biol. **59**(4), R151–R182, (2014). [*PMC*. Web. 21 (2018)]
6. D. Inman, R. Elmore, B. Bush, A survey onVLSI architectures of lifting based 2D discrete wavelet transform. Build. Serv. Eng. Res. Technol. **36**(5), 628–637 (2015). https://doi.org/10.1177/0143624415573215
7. C. Beccari, G. Casciola, L. Romani, Computation and modeling in piecewise Chebyshevian spline spaces. Numer. Anal., arXiv:1611.02068, (2016)

8. A. Grigorenko, S. Yaremchenko, Investigation of static and dynamic behavior of anisotropic inhomogeneous shallow shells by Spline approximation method. J. Civ. Eng. Manag. **15**(1), 87–93 (2009). https://doi.org/10.3846/1392-3730.2009.15.87-93
9. C-G. Zhu, X-Y. Zhao, Self-intersections of rational Bézier curves. Graph. Model. **76**(5) (September 2014), 312–320, (2014). http://dx.doi.org/10.1016/j.gmod.2014.04.001

Chapter 4
Multidimensional Polynomial Splines

4.1 Multidimensional Polynomial Splines in Modelling of Functional Relationships

During the process of scientific research and test work, automated processing the measuring results in computing systems and complexes, due to difficulties in the unified analytical description of large size functional relationships, an approach is naturally developed that uses segmentation of areas into segments and piecewise description of signal models [1]. The problems with identification of real relationships are complicated if the information is not complete, heavily deviated by disturbances, or the functions are of multi-extreme nature, etc.

The piecewise-polynomial and piecewise-rational methods are simplest and are implemented fast and efficiently by hardware computer means [2–4]. But the classical bracket expression of polynomial models $P_m(x)$ starting from Horner's algorithm for a one-dimensional case [5]:

$$P_m(x) = ((\ldots (a_m x + a_{m-1})x + a_{m-2})x + \ldots + a_1)x + a_0, \qquad (4.1.1)$$

indicates the complication of parallelising of calculations.

Subject to the h argument step—distance between nodes—the interpolation polynomial splines have better approximation convergence assessments in comparison with interpolation polynomials [6]. So, for a quadratic spline, the following inequality is true:

$$\varepsilon \leq (\sqrt{3/216})\max\left|(f^{(3)}(x)\big|h^3\right) \qquad (4.1.2)$$

where f—function for interpolation, ε—interpolation inaccuracy.

For a cubic spline, the inequality looks like the following:

$$\varepsilon \leq (a/384)\max\left|f^{(4)}(x)\big|h^4\right) \qquad (4.1.3)$$

© The Author(s), under exclusive license to Springer Nature Singapore Pte Ltd. 2019
D. Singh et al., *Signal Processing Applications Using Multidimensional Polynomial Splines*, SpringerBriefs in Applied Sciences and Technology,
https://doi.org/10.1007/978-981-13-2239-6_4

where the constant "a" takes values from 1 to 5, while the lower limit coincides with Hermite's splines.

At computer solution of interpolation tasks with splines, all polynomial specific parallelising issues still exist. In addition, the spline coefficient memory takes large volumes because the spline requires $m + 1$ coefficient for an interval of the argument between two adjacent nodes, while a polynomial requires $(m + 1)/m$ at an average [7].

The depth of pipeline structures that calculate polynomial splines and that use summator sets, multipliers and multiport memory coefficients, increases at an increase of the level of the spline, which may lead to significant delays in obtaining a result.

The function calculation parallelising issues with several variables are significantly complicated upon increase of the level of approximating polynomials on each argument. The known theoretical algebraic model [8] where the coefficients at the level of one argument are expressed as polynomials on another argument

$$f(x, y) = \sum a_i(x) y^i \tag{4.1.4}$$

$$a_i(x) = \sum c_k x^k \tag{4.1.5}$$

leads to the case, when in comparison with a one-dimensional model, the number of algorithm steps considerably rises.

The empiric multidimensional functional relationships, obtained as a result of experimental studies, are significantly processed, while being represented by piecewise surfaces of lower levels due to measurement inaccuracies and high disturbance levels.

So, owing to the universality of readout processing algorithms, well differential and extreme properties, high convergence of approximation assessments, simplicity of calculation of forms and parameters and poor interference of rounding errors, splines as a class of piecewise functions are more broadly used in construction of hardware and software means for analysis and recovery of one-dimensional and multidimensional signals, thus expanding the range of traditional approaches.

4.2 Methods of Computation of Modelling Coefficients Using Multidimensional Splines

The theory of approximation of a function of many variables [9] with multidimensional splines has lately seen significant development. If the interpolating polynomial splines area is taken, all the determination of one-dimensional splines is naturally expanded for the cases with several arguments [10]. So, the function $S_m(x, y)$ is called a spline with two variable of the level m relative to the net $\{x_i, y_i\}$, if it coincides with the polynomial of level m on x and on y at each rectangle D.

For any y from D, a one-dimensional spline can be constructed that interpolates the values vector $P(y)$ on the net. $S_m(P(y), x)$ is called a partial spline. Its use in the construction of a two-dimensional spline consists of a combination of one-dimensional splines:

$$S_m(x, y) = \sum_{j=0}^{n} P_{ij}(x) S_j(y) \tag{4.2.1}$$

where $P_{ij}(x)$—polynomials of level m on x, while $S_j(y)$—a polynomial of level m on y at $x, y \in$ D.

Multidimensional polynomial B-splines of equal level m on each argument are determined as tensor direct products of one-dimensional B-splines [11]:

$$B_m(x, y, \ldots, u) = B_m(x) \otimes B_m(y) \otimes \cdots \otimes B_m(u) \tag{4.2.2}$$

In particular, for a two-dimensional spline $S_m(x, y)$ of level m, the following formula applies:

$$S_m(x, y) = \sum \sum b_{ij} B_{m,i}(x) B_{m,j}(y), \tag{4.2.3}$$

i.e., as dual sums of multiple products, where the multipliers are the coefficients and one-dimensional B-splines. Here, the determination area of nonzero values of the two-dimensional basic spline

$$B(x, y) = B(x) \otimes B(y) \tag{4.2.4}$$

represents a rectangle $[x_i, x_{i+1}; y_j, y_{j+1}]$ produced from the net segmentation of the following type:

$$\Delta x : x_0 < x_1 < x_2 < \cdots < x_{n1-1} < x_{n1}; \tag{4.2.5}$$

$$\Delta y : y_0 < y_1 < y_2 < \cdots < y_{n2-1} < y_{n2}. \tag{4.2.6}$$

A successive-parallel algorithm of computation of multidimensional spline values can be implemented, if the inter-dependent values are entered for coefficients [12]. In particular, for the case of two independent arguments, we have:

$$c_i(y) = \sum_j b_{ij} B_j(y), \tag{4.2.7}$$

and the expression for the spline would be recorded as:

$$S_m(x, y) = \sum_i c_i(y) B_i(x), \tag{4.2.8}$$

i.e., it can be computed through pair products in two stages.
Similarly, the three-dimensional spline is represented as

$$S_m(x, y, z) = \sum_i \sum_j \sum_k b_{ijk} B_i(x) b_j(y) B_k(z) \qquad (4.2.9)$$

and its values can be produced as a result of computing successions of pair products.

Local coefficient computation formulae can also be constructed for multidimensional approximation cases. For instance, the three-point formula for a bi-cubic spline at uniform/equal nets Δx and Δy can be obtained on the basis of the formulae for one-dimensional splines [8]:

$$a_{1j} = 6f_{kj} - 4a_{kj} - a_{k-1j}, \qquad k = n_1 - 1, n; \quad j = 0, 1, \ldots, n_2;$$

$$b_{ij} = (-a_{i,j-1} + 8a_{ij} - a_{i,j+1})/6, \quad j = 1, 2, \ldots, n_2 - 1;$$

$$b_{i,k-1} = 6a_{ik} - 4b_{ik} - b_{i,k+1}, \qquad k = 1, 0;$$

$$a_{ij} = (f_{i-1j} + 8f_{ij} - f_{i+1j})/6, \qquad i = 1, 2, \ldots, n_1 - 1;$$

$$a_{k-1j} = 6f_{kj} - 4a_{kj} - a_{k+1j}, \qquad k = 1, 0;$$

$$a_{k+1j} = 6f_{kj} - 4a_{kj} - a_{k-1j}, \qquad k = n_2 - 1; \quad i = -1, 0, \ldots, n_1 + 1.$$

$$(4.2.10)$$

Therefore, the local feature of one-dimensional splines is entirely applied to the multidimensional splines. At a similar approximation step, the two-dimensional spline can be represented as two one-dimensional splines.

4.3 The Main Characteristics of Implementation of Tabular-Algorithmic Computation Structures of Piecewise-Polynomial Processing of Signals

4.3.1 Performance Assessment

The time for performing an approximation at a random point for a tabular-algorithmic computation structure on the basis of classical polynomials is determined using the formula [9]:

$$t_t = t_{four} + t_{mult} + t_{sum} + t_{mult} + t_{sum} = t_{four} + 2t_{four} + 2t_{sum}.$$

Inserting relevant values, we shall produce:

$$t_t == t + 2t_{mult} + 2t_{sum} = 60 + 290 + 100 = 450 \text{ ns},$$

where t_{mult}—the time of multiplication, the value of which for series K1802 equals $t_{mult} = 145$ ns.

The time for performing an approximation on a random point for a tabular-algorithmic structure on the basis of parabolic basic splines is determined using the formula:

$$t_t = t_{four} + t_{mult} + t_{sum} = 60 + 145 + 50 = 255 \text{ ns.}$$

The time for performing an approximation on a random point for a tabular-algorithmic structure on the basis of the third-level classical polynomials is determined using the formula:

$$t_t = t_{four} + 3t_{mult} + t_{sum}.$$

Or by inserting relevant values, we shall produce:

$$t_t = t_{four} + t_{mult} + t_{sum} = 60 + 145 + 50 = 255 \text{ ns.}$$

The time for performing an approximation on a random point for a tabular-algorithmic structure on the basis of cubic basic splines is determined using the formula:

$$t_t = t_{four} + 3t_{mult} + 3t_{sum} = 60 + 3 \cdot 145 + 3 \cdot 50 = 645 \text{ ns.}$$

4.3.2 Memory Capacity Assessment

The tabular-algorithmic computation structures require memory storage capacity of 1 Kb, buffer registers and a basic function memory storage (BFMS). Tables 4.1 and 4.2 show the main characteristics of tabular-algorithmic high-performance computation structures of piecewise-polynomial processing of signals.

Accordingly, the main characteristics of the implementation of tabular-algorithmic computation structures of piecewise-polynomial processing of functions were obtained. They allow making the following conclusions:

1. The computation structures, obtained on the basis of basic splines, function (the second level—1.76 times, and for the third level—2.53 times) faster in comparison with the structures, obtained on the basis of the same levels' classical polynomials.
2. The time for performing an approximation at a point for parabolic and cubic basic splines is equal 255 ns, while the cubic basic splines require more hardware costs (one summator and one multiplier more).

Table 4.1 Characteristics of tabular-algorithmic high-performance computation structures on the basis of classical polynomials and the third-level basic splines

No.	Description	Structure on the basis of the third-level classical polynomials	Structure on the basis of cubic basic splines
1	Time for performing an approximation on a point	645 ns	255 ns
2	Number of buffer registers	–	5 pcs
3	Number of memory capacity	1 pcs	1 pcs
4	Number of summators	3 pcs	1 pcs
5	Number of multipliers	3 pcs	4 pcs
6	BFMS	–	1 pcs

Table 4.2 Characteristics of implementation of tabular-algorithmic high-performance computation structure on the basis of classical polynomials and the second-level basic splines

No.	Description	Structure on the basis of the second-level classical polynomials	Structure on the basis of parabolic basic splines
1	Time for performing an approximation at a point	450 ns	255 ns
2	Number of buffer registers	–	4 pcs
3	Number of memory capacity	1 pcs	1 pcs
4	Memory capacity	1 Kb	1 Kb
4	Number of summators	2 pcs	1 pcs
5	Number of multipliers	2 pcs	3 pcs
6	BFMS	–	1 pcs

4.4 Summary

This chapter has described the combination of advantages of the tabular-algorithmic methods for reproduction of functions and multidimensional B-splines which leads to the implementation of parallel-pipeline computation structures of multiple variables that ensure the highest performance.

Some multiprocessor high-performance computation structures were evaluated, based on the use of the tabular-algorithmic method of processing and multidimensional spline approximation that are noted for high performance, actually top speed for tabular-algorithmic methods. A parallel-pipeline computation structure has been developed for implementation of two-dimensional basic spline approximation. It allows saving the memory for storage of values of basic splines twice at a limited number of processors. A parallel-pipeline computation structure is proposed for recovery of the value of three-variable functions that have a limited number of processors and are noted for high performance.

Some main characteristics of the implementation of tabular-algorithmic computation structures were obtained for processing of signals in piecewise-polynomial basics. It was proven that the tabular-algorithmic computation structures on the basis of basic splines function faster (in the case of parabolic basic splines 1.76 times and in the case of cubic basic splines—2.53 times) than the classical polynomials of the same levels. It has also been proven that the hardware costs for implementation of the computation structures based on cubic splines are higher at the same time costing.

References

1. R.Z. Morawski, On teaching measurement applications of digital signal processing. Measurement **40**(2), 213–223, ISSN 0263-2241, (2007). https://doi.org/10.1016/j.measurement.2006.06.015
2. A. Sotiras, C. Davatzikos, N. Paragios, Deformable Medical image registration: a survey. IEEE Trans. Med. Imaging **32**(7), 1153–1190, (2013). https://doi.org/10.1109/tmi.2013.2265603. (*PMC*. Web. 21 Sept. 2018)
3. I. Garcia Marco, P. Koiran, T. Pecatte, Polynomial equivalence problems for sum of affine powers. PRoceddInt3rpp 303–310, (2018). https://doi.org/10.1145/3208976.3208993
4. V. Agrawal, K. Bhattacharya, Shock wave propagation through a model one dimensional heterogeneous medium. Int. J. Solids Struct. **51**(21–22), 3604–3618, ISSN 0020-7683, (2014). https://doi.org/10.1016/j.ijsolstr.2014.06.021
5. I.V. Anikin, K. Alnajjar, Primitive polynomial selection method for peseudo-random number generator. J. Phys.: Conf. Ser. **944**(1), 012003, (2018). http://stacks.iop.org/1742-6596/944/i=1/a=012003
6. D. Botes, P.M. Bokov, Polynomial interpolation of few-group neutron cross sections on sparse grids. Ann. Nucl. Energy **64**, 156–168, ISSN 0306-4549, (2014)
7. G. Raspa, F. Salvi, G. Torri, Probability mapping of indoor radon-prone areas using disjunctive kriging. Radiat. Prot. Dosimetry **138**(1), 3–19 (2010). https://doi.org/10.1093/rpd/ncp180
8. T. Hämäläinen, J. Saarinen, K. Kaski, TUTNC: a general purpose parallel computer for neural network computations. Microprocess. Microsyst. **19**(8), 447–465, ISSN 0141-9331, (1995). https://doi.org/10.1016/0141-9331(96)82010-2
9. M. Viegers, E. Brunner, O. Soloviev, C.C. de Visser, M. Verhaegen, Nonlinear spline wavefront reconstruction through moment-based Shack-Hartmann sensor measurements. Opt. Express **25**(10), 11514–11529 (2017). https://doi.org/10.1364/OE.25.011514
10. Y.-M. Wang, L. Ren, High-order compact difference methods for caputo-type variable coefficient fractional sub-diffusion equations in consecutive form. J. Sci. Comput. **76**(2), 1007–1043 (2018). https://doi.org/10.1007/s10915-018-0647-4
11. T.A.M. Langlands, B.I. Henry, The accuracy and stability of an implicit solution method for the fractional diffusion equation. J. Comput. Phys. **205**(2), 719–736 (2005). http://dx.doi.org/10.1016/j.jcp.2004.11.025
12. Z.Q. Li, Z.Q. Liang, Y.B. Yan, High-order numerical methods for solving time fractional partial differential equations. J. Sci. Comput. **71**, 785–803 (2017)

Chapter 5
Evaluation Methods of Spline

5.1 Hardware Used for Evaluation of Spline-Recovery Methods

In computer technologies, the arithmetic processing of a succession of readouts that are equidistant in time is called digital processing of signals. The algorithm of digital processing is also used for computation over massifs for the determination of data correlation and two-dimensional filtrations [1–4].

As opposed to analogue processing, which is traditionally used in many radio technical devices, digital processing is a cheaper method of achieving the result and ensures higher accuracy, diminutiveness and fabricability of gadgets, the flexibility of realignment and thermal stability.

It is out of the question that digital processing of a signal can be performed using usual computation hardware. For instance, on a modern personal computer with a Pentium IV processor with two computation cores and possibility of pipeline working, computations are not difficult. However, within the strict real-time limitation system, for which the digital processing systems were evaluated initially, the very specifics of successiveness offer additional opportunities for achieving high performance efficiency.

The spline-function methods are convenient to the possibility of broad use of parallelising principles, a combination of data entry with processing. Signal recovery using spline-function methods is based upon the performance of operations of parallel multiplication with accumulation.

The presence of parallel multipliers with accumulation, a data storage unit with free access and software evaluation instruments in digital processors of signals allows their use in signal processing tasks using spline-function methods.

The modern use of digital processing methods is considered to be in the multimedia technologies field, i.e. sound and image processing that includes their compression and coding. In digital communication field, digital methods perform the modulation and demodulation for transfer along communication channels.

The radiolocation systems put stricter requirements to a hardware component of digital processing. Here, the main content of digital processing is filtration of entry signals of the antenna, signal frequencies from 10 MHz to 10 GHz and sizes of transformations as high as 2^{14} complex points, and the requirement to performance may be as high as 10^9 multiplications a second. Processing of radiolocator digital signals utilises algorithms of digital filtration and spectral analysis, algorithms of correlational analysis, reverse compression and special algorithms for linear prediction.

In sound processing systems, digital processors of signal processing deal with analysis, recognition and synthesis of speech and compression of speech in telecommunication systems. Processing of digitised sound signals uses algorithms of digital filtration and spectral analysis, algorithms of correlational analysis, reverse compression and special algorithms for linear prediction. In most cases, satisfactory results are ensured by the data format with fixed comma, the word length of 16 bytes, signal frequency from 4 to 20 kHz (to 40 kHz in music processing), required performance—to 10×10^6 operations a second—10 MIPS as per computer terminology.

For image processing systems, image enhancement, data compression for transfer and storage and image recognition are typical tasks. Recovery/restoration and enhancement of images using inverse compression and massive readout processing using Fourier's fast transformation algorithms are specific tasks for image processing systems. The spatial-frequency filtration methods are used for recovery of three-dimensional objects that are obtained using the methods of penetrating radiation in defect detection and medical imaging processes.

Another class of algorithms is the transformation of contrast range, contour extraction and statistic processing of images. Fourier's, Adamar's and Walsh's orthogonal transformations are most efficient for data compression. The required performance rate is assessed by the value range of 100–1000 MIPS, data massifs—10^5–10^6 readouts.

A signal is a function pattern (image) of change of a physical parameter of type, for instance, temperature, pressure, tension or brightness. In the digital processing theory, a signal is the dependence of tension on time.

The initial physical signal is a continuous function of time. Such signals, determined in all moments of time, are called analogue signals. The successiveness of numbers that represents a signal in digital processing is a discrete row and cannot completely correspond to the analogue signal. The numbers that compose successiveness are the values of the signal in discrete moments of time and are called signal readouts. As a rule, the readouts are taken at equal time intervals T, which are called discretisation periods. The value reciprocal to the discretisation period is called discretisation frequency.

5.2 Purpose and Characteristics of Signals

The process of transformation of an analogue signal into successive readouts is called discretisation, while the result of such transformation is called a discrete signal. Table 5.1 contains the types, purpose and characteristics of signals.

Table 5.1 Fields of use, purpose and characteristics of signals

Name	Purpose	Characteristics of signals
Radio- and hydro location	Discovery and measurement of coordinates, profiling, construction of radio image	Signal band 100 kHz–1 MHz, signal base 10^2–10^4
Radio communications	Ensuring the reliability of surface and cosmic communications owing to optimisation of allotment algorithms, coding for compression and enhancement of interference immunity, interference suppression	
Radio astronomy	Line spectres, radio interferometry, with super-large bases for resolution of radio sources	Signal band 1–10 MHz
Image processing	Enhancement of image quality, sharpness, and contrast, interference suppression, compression and recovery after compression, image recovery from digitised holograms	Image sizes $10^3 \times 10^3$
Speech and sound signal processing	Speech analysis and synthesis, enhancement of quality of sound recording, of acoustics of premises and systems	Signal band to 50 kHz
Geophysics	Analysis of natural seismic signals for control and prediction of earthquakes, processing of results of seismic exploration for description of geological structures under surface	LF >300 kHz at 10 mcs length, HF = 10–50 kHz at 100–400 mcs length
Medicine, biology	Analysis of cardiograms and encephalograms, tomographic scanning, animal sound analysis	Signal band to 400 kHz
Vibration analysis	For control of engine and mechanical systems quality	Signal band 100 kHz–1 MHz

5.3   Structure of DSP with Harvard's Architecture

Processors for digital processing of signals (DSP) are in much broader use in many different fields for they are capable of ensuring the work in a real time of both: the existing and principally new gadgets.

A digital signal processor is a microprocessor with a specific work mode of stream processing of large scopes of data in real time and, as a rule, with intensive data exchange with external gadgets [5].

A real-time scale is a working mode of a gadget, when registration and arithmetic processing (if required, analysis, visualisation, storage, systematisation, synthesis and transfer along communication channels) of data is performed without any loss of the information that reports from its sources [6–9].

New hardware is developed on the basis of DSP for performing arithmetic tasks, while we can identify a number of tasks using that are most frequently performed using DSP [10]:

- Signal filtration;
- Compression of two signals (mixing of two signals);
- Identification of values of auto- and cross-correlational functions of two signals;
- Enhancement, normalisation and transformation of signals;
- Direct and reciprocal Fourier's transformation;
- And others.

The specific feature of the digital signal processing tasks is the stream processing of large scopes of data in a real-time scale that requires high performance rates from hardware and the possibility of intensive data exchange with external hardware. This is achieved owing to the specific structure of DSP, which is called the basic architecture of DSP.

The basic architecture of DSP is the combination of typical features of the processor, intended for the enhancement of its performance rate, and that differentiates DSP from other type microchips, and it is explained by [11]:

- The use of modified Harvard architecture;
- Broad use of conveyor work mode;
- Existence of a specialised multiplication unit;
- Existence of special commands for digital processing of signals;
- Implementation of sort command cycle.

The main feature of the Harvard architecture is the use of separate spaces for storage of commands and data (Fig. 5.1).

An analysis of real management software showed that the necessary scope of DSP data, used for storage of intermediate results, as a rule, is a lot smaller than the required scope of software memory. In such conditions, the use of a unified address space leads to an expansion of command formats due to an increase in the number of charges for addressing of operands. The use of a separate small volume memory of data facilitates reduction of the length of commands and acceleration of information search in data storage.

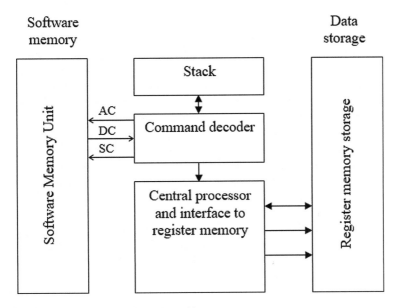

Fig. 5.1 Structure of DSP with Harvard's architecture

In addition, Harvard's architecture ensures potentially higher software performance velocity in comparison with other software owing to the possibility of implementation of parallel operations. Instruction of the next command can occur simultaneously with performance of the previous command; there is no necessity in stopping the processor for command instruction time. This method of implementation of operation allows ensuring the performance of different commands during the same number of time periods, which gives an opportunity in the simpler determination of the time of cycle performance and critical areas of the software.

Most modern DSP manufacturers use the Harvard architecture. However, Harvard's architecture is not sufficiently flexible for the implementation of software procedures. Therefore, the comparison of DSP, performed on different architectures, should be undertaken in application to specific tasks.

An analysis of market shows that large companies, like Texas Instruments Inc., Analog Devices, Motorola and a number of others, take leading positions in the production of DPSP. They are capable of not only creating new low-cost DPSP, but also annually investing in new evaluations and know-how and creating principally new models and platforms.

Products of the following companies can be named as most widely used—Texas Instruments Inc. (TMS320), Motorola (DSP56F800) and Analog Devices (ADSP-BF5xx). Selection of processors of a company for the purposes of a specific project is a multiple component tasks, and it is actually impossible to formulate a more or less clear methodology of selection. Therefore, we shall discuss DPSP of several companies.

DSP TMS320C2x three series—TMS320C2000, TMS320C5000 and TMS320C6000, according to the experts of the company, in the near future—are intending to cover the whole range of possible usages of DSP's, offering the consumers a selection of DSP on "performance/cost/used power" criteria [12].

The TMS320C2000 DSP series is intended for the solution of built-in application tasks and management [12]; the processors are noted for developed periphery and low cost; these series are represented by universal DSP subseries TMS320C20x and subseries TMS320C24x for digital management of electric motors.

DSP series TMS320C5000 is oriented for low-power consumer markets of potable gadgets and mobile communications. The DSP of series TMS320C54xx is optimised on performance rate (to 200 MIPS) and minimum power consumption (to 32 mA/MIPS). In addition, overall use of 0.18 mcm technology enabled the cost reduction of certain DSP of this subseries to 5$ at 100 MIPS performance rate.

The DSP series TMS320C6000 is characterised by maximum performance rate for applications that require top speeds of computations with fixed and floating points. The standard fields of use of DSP series TMS320C6000—multichannel modems, base stations, image processing hardware and others.

All three DSP series can be equipped with modern processing and software debugging tools, combined with one user interface based on Code Explorer and Code Composer Studio software.

The DSP56600 family, intended for mobile wireless communications, implemented the idea of combining two cores: 32-bit RISC core MCORE and DSP core on one crystal. The use of such architecture required an evaluation of separate programs for each core and their further combined debugging, which was a labour-consuming process.

To date, Motorola stopped further support of the DSP56600 family and concentrated its efforts on the development of the DSP56F800 family. The processors of the latter are produced by hybrid technology of 0.25 mcm, and the company engineers have a target of combining the DSP performance rate and functionality of microcontroller in one microchip.

The DSP56F800 does not contain two processor cores, placed in one box. And it requires from designer's development of two different software types, but it represents a new core that has both: the DSP and microcontroller features. Its core is capable of performing management tasks and fast DSP algorithms on data processing. The necessity of working on one software only significantly simplifies the process of evaluation of the application.

The 16-bit DSP core was chosen as the base for DSP56F800. This core is the development of the Motorola DSP56100 architecture. Instructions for working with integer numbers, oriented for management tasks, were included; also, a one-bit processor was introduced. The multiplicity of DSP56F800 addressing modes probably reminds a microcontroller than a DSP. For support of efficient software evaluation on high-level language, a register-stack indicator, specific for microcontrollers, was added into the architecture. This register allows organising the software stack and has unlimited depth of nesting of subprograms.

Performance and functionality, availability of a flash memory, simplicity and ease of software development in combination with low cost guaranteed the DSP56F800 a wide range of possible applications.

5.4 Blackfin Family Processors

The DSP of Blackfin family of Analog Devices such as the ADSP-BF531, ADSP-BF532 and ADSP-BF533 processors are representatives of the Blackfin processor family with advanced facilities that have significantly higher performance rate and less power consumption in comparison with previous processors of Blackfin family while keeping the simplicity of use and code compatibility [13]. Three new processors are completely compatible on outputs and differ from each on performance rate and internal memory sizes, which enables avoidance of difficulties that occur in the development of new products.

The Blackfin processor core architecture is architecture with a unified command set that includes a signal processing core with a double block for multiplication-accumulation, which has orthogonal command set, specific for RISC micropro-cessors, and which has a flexibility of SIMD type commands and multimedia possibilities.

The specific feature of Blackfin family products is the dynamic management of power supply. The possibility of changing both, the supply voltage and working frequency, enables optimisation of power supply according to specific tasks.

All periphery units save the general-purpose input/output port, real-time clock and timers and are supported by a structure of flexible direct access to memory (DAM). The processor also has two separate DAM channels of "memory–memory" type, intended for implementation of transfers between processor memory spaces, including external SDRAM and asynchronous memory. The set of internal buses ensures the memory bandwidth, sufficient for supporting the processor core, even if all internal and external periphery gadgets are involved.

The processor core contains two 16-bit multipliers, two 40-bit accumulators, two 40-bit arithmetic logic units (ALUs), four 8-bit video ALU and a 40-bit displacement unit [14]. The computation units process the 8-, 16- or 32-bit data that report from the register file.

The computation register file contains eight 32-bit registries. At performing com-putation operations on 16-bit operands, the registry file function has 16 independent 16-bit registries. All the operands of computation operations report from the multi-port registry file or are instructed by constant values in the command fields.

At one-time period, each multiplier-accumulator (MAC) performs multiplication of two 16-bit numbers and accumulation, thus forming a 40-bit result. The signed and unsigned number formats, rounding and saturation are supported.

ALU performs a traditional set of arithmetic and logical operation on 16- or 32-bit data. It includes many special commands that accelerate the performance of various signal processing tasks. These tasks include bit operations like extraction of the field, calculation of the number of units, multiplication on module 2^{32}, primitives of division, saturation and rounding and determination sign/order. The set of video commands includes the operation of levelling and packaging of bites, adding of 16-bit numbers with 8-bit numbers with reduction/truncation of the result, operations of 8-bit averaging and operation of 8-bit deduction, of taking the absolute value, accumulation (SAA, subtract, absolute value, accumulation). Commands of comparison/selection and vector search are also supported. At the use of some commands, simultaneous performance of two 16-bit ALU operations is possible over the registry pairs (small and large 16-bit halves of computation registry). At the use of a second ALU, simultaneous performance of four 16-bit operations is possible.

The 40-bit shifter unit can perform data entry and shifting operation, of cycling shift, normalisation and extraction of the bit field.

The programmed automation manages the command performance process, including the performance of command levelling and decoding operations. During management of software performance, the software automates supports relative (to command registrar) and indirect conditional conversions (on the statistical indication of conversions) and calls of subprograms. The processor has an implemented hardware support of cycles with zero nonproductive costs. The programmed automation architecture is completely closed, which ensures the lack of visible effects of the work of conveyor while performing the commands with interrelated data.

The address arithmetic unit forms two addresses for performing simultaneous dual sets from memory. It contains a multiport registry file that consists of four sets of 32-bit index registers, modification, length and a basic address (for the organisation of cyclic buffers) and eight additional 32-bit indicator registries (for indexed manipulation of the stack in C language style).

The Blackfin processors support modified Harvard architecture with a hierarchic memory structure. The Level 1 memory (L1) usually works at the full speed of the processor with a little delay or without any delay. The command memory at L1 level contains commands only. Two data memories contain the data, while the allotted super-operative memory stores the information of the stack and local variables [15].

The processor has several L1 memory blocks, which may be configured as a mixed set of SRAM and cache. The memory management unit (MMU) ensures the protection of memory, while the core performs individual tasks and can protect the system registers from unintentional access.

The processor architecture offers three work modes: user's mode, supervisor's mode and emulation mode. At user's mode, access to system subset of resources is limited, which ensures organisation of protected software environment. In supervisor and emulation modes, access to the core and system resources is not limited.

The command set of ADSP-BF53x Blackfin processor is optimised in a way that the most frequently used commands are represented by 16-bit codes. The complex commands of digital processing of signals (DPS) are coded with 32-bit codes as multifunctional commands. The limited support for multitasking is implemented in Blackfin family products. Multitasking offers the possibility of parallel call of a 32-bit command and two 16-bit commands. This enables designers to use the many resources of the core in one command cycle.

The ADSP-BF53x Blackfin processor assembler language uses an algebraic syntax. The architecture is optimised for joint use with the C language compiler.

The Blackfin processor architecture memory structure has a unified 4-Gb address space that uses 32-bit addresses. All resources, including the internal memory, external memory and input/output management registries take a separate section in general address space. The address space storage areas are sorted in a hierarchic structure, which ensures the cost and performance balance in the use of very fast internal memory, like cache or SRAM that has a very little delay and low-cost and high-performance external systems of large volumes. Tables 5.2 and 5.3 contain comparisons of ADSP-BF531, ADSP-BF532 and ADSP-BF533 processors.

Blackfin offers a new-generation 16/32-bit built-in processor with high performance characteristics and efficient power for applications, with similar possibilities: multiformat audio, video, sound and processing of images; multimode bandwidth of modulating frequencies and package processing; and real-time protection and processing management are critical. This very powerful combination of software

Table 5.2 Comparison of memory of Blackfin family processors

Memory type	DSP		
	ADSP-BF531	ADSP-BF532	ADSP-BF533
SRAM/cache command	16 Kbite	16 Kbite	16 Kbite
SRAM commands	16 Kbite	32 Kbite	64 Kbite
MU commands	32 Kbite	32 Kbite	–
SRAM/cache of data	16 Kbite	32 Kbite	32 Kbite
SRAM of data	–	–	32 Kbite
Super-operative memory	4 Kbite	4 Kbite	4 Kbite
Total memory capacity	84 Kbite	116 Kbite	148 Kbite

Table 5.3 Comparative characteristics of Blackfin family processors

DSP name	Maximum frequency at a time, MHz	Maximum performance rate (MMACS)	Internal crystal memory, Kb
ADSPBF531	400	800	52
ADSPBF532	400	800	84
ADSPBF533	756	1512	148
ADSPBF534	500	1000	132
ADSPBF535	350	700	308
ADSPBF536	400	800	100
ADSPBF537	600	1200	132

flexibility and universality was widely distributed in the convergent application of the type of digital domestic entertainment; in networks and communications; automatic integrated information processing means and data transfer; and digital radio and mobile television. The main features of the processor are:

- An architecture with a set of one-stream commands with processing speed is offered or wins the competition in the range of DSP products—ensures the best speed, cost and memory efficiency.
- 16/32-bit architecture allows next-generation built-in applications and systems.
- Management, signal and multimedia processing in one core.
- Performance rate that can be aligned for processing of signals or consumed power via management of dynamic power.
- Code portfolio and connector compatible product. The cost under 5$ to 1.500 MIPS influences all final products developments.
- It improves the designer production rate.
- Minimum optimisation is required owing to the powerful programming environment, associated with core performance rate.

The Blackfin core which is a 32-bit RISC processor with frequencies from 400 to 750 MHz is the heart of the Analog Devices platforms. It is a powerful, cost-efficient, well-optimised tool for the processing of video. An additional advantage is easy adaptability of codes of large numbers of respective solutions and therefore the easy implementation of various programme components.

The Blackfin core demonstrates good cost-saving features: from a 600 to 800 mA/h battery, from 5 to 9 h of QVGA resolution video and to 16 h of music can be played.

Digital signal processors of Blackfin ADSPBF534, ADSPBF536 and ADSPBF537 family are designed for highly intellectual gadgets that have network interfaces ensuring high network safety. These processors have been widely used for all types of multimedia IT applications, industrial controllers with network protocol support, intellectual on-board car/truck systems and many other applications, which require a network interface, little power supply and low cost.

The Blackfin family processors combine digital signal processor capabilities of efficient audio and video information processing with flexibility and simplicity of use of 16/32-bit microcontrollers, which expands the use area of these processors to the fields, where general-purpose microcontrollers have been used. The combination of network interfaces and a high-rate processor core allows the engineers evaluate new gadgets with less cost, larger integration and new capabilities in comparison with the previous generation gadgets.

All Blackfin processors are fully supported by evaluation instruments of Analog Devices CROSSCORE that include an integrated evaluation environment VisualDSP++, EZKIT debugging sets, EZExtender expansion patches to debugging sets and emulators. Easy-to-use VisualDSP++-integrated evaluation environment allows time saving for software evaluation and debugging. VisualDSP++ 4.0 supports TCP/IP and USB protocols and contains processor configuration and initialisation master.

Therefore, Balckfin family processors are most applicable among the described DSP for processing of the selected classes of signals. The cost of these processors is relatively cheap, and performance rate and required storage capacity in these processors meet the requirements of signal processing of both: the results of bench tests and geophysical signals.

5.5 Summary

The studies of the spline-function methods indicate that as in the algorithmic computation of coefficients as well as in algorithms of recovery of signals there are frequent operations like parallel additions, multiplication and multiplication with accumulation. These operations are typical for digital processing of signals. Therefore, the algorithms, obtained because of application of the spline-function methods, are convenient for the implementation of use in digital signal processors.

An analysis of existing hardwares are intended for digital processing of signals indicated that Harvard's architecture and availability of hardware implemented special multiplication commands, parallel multiplication with accumulation offer an opportunity of wide use of modern digital signal processors for implementation of spline-recovery methods.

Some instrumental tools of modelling of the processes of recovery of one-dimensional and multidimensional signals for digital signal processors of Blackfin family of Analog Devices have been proposed. The instrument allows the designers develop and perform debugging of applications. This environment includes an easy-to-use assembler. A key feature of software development tools is the efficiency of the code that is written in C/C++ languages.

A class of digital signal processors has been analysed, and the Analog Devices ADSP Blackfin processors were selected for the implementation of spline-recovery methods. This processor family best fits for processing of the chosen class of signals. The cost of such processors are relatively low; and have parallel multipliers with accumulation and dynamic management of power source, performance rate and the required volume of memory storage capacity for processing of signals and supports integrated VisualDSP++ development environment.

References

1. Y.-M. Wnag, L. Ren, G. Ao, Z. Sun, H. Zhang, A new fractional numerical differentiation formula to approximate the Caputo fractional derivative and its applications. J. Comput. Phys. **259**, 33–50 (2014)
2. A.A. Alikhanov, A new difference scheme for the time fractional diffusion equation. J. Comput. Phys. **280**, 424–438 (2015)
3. H. Zhang, X. Yang, D. Xu, A high-order numerical method for solving the 2D fourth order reaction-diffusion equation. Numer. Algorithms 1–29 (2018)

4. C.P. Li, R.F. Wu, H.F. Ding, High-order approximation to Caputo derivative and Caputo-type advection-diffusion equations. Commun. Appl. Ind. Math. **6**(2), e-536 (2014)
5. H. Li, J. Cao, C. Li, High-order approximation to Caputo derivatives and Caputo-type advection-diffusion equations (III). J. Comput. Appl. Math. **299**, 159–175 (2016)
6. C. Lv, C. Xu, Error analysis of a high order method for time-fractional diffusion equations. SIAM J. Sci. Comput. **38**, A2699–A2724 (2016)
7. Z.Q. Li, Y.B. Yan, N.J. Ford, Error estimates of a high order numerical method for solving linear fractional differential equations. Appl. Numer. Math. **114**, 201–220 (2016)
8. Y.B. Yan, K. Pal, N.J. Ford, Higher order numerical methods for solving fractional differential equations. BIT Numer. Math. **54**, 555–584 (2014)
9. M. Dehghan, M. Abbaszadeh, Two meshless procedures: moving Kriging interpolation and element-free Galerkin for fractional PDEs. Appl. Anal. **96**, 936–969 (2017)
10. M. Dehghan, M. Abbaszadeh, Element free Galerkin approach based on the reproducing kernel particle method for solving 2D fractional Tricomi-type equation with Robin boundary condition. Comput. Math. Appl. **73**, 1270–1285 (2017)
11. X. Yang, H. Zhang, D. Xu, J. Sci. Comput. (2018). https://doi.org/10.1007/s10915-018-0672-3
12. B. Jin, R. Lazarov, J. Pasciak, Z. Zhou, Error analysis of semidiscrete finite element methods for inhomogeneous time-fractional diffusion. IMA J. Numer. Anal. **35**, 561–582 (2015)
13. B. Jin, R. Lazarov, Y. Liu, Z. Zhou, The Galerkin finite element method for a multi-term time-fractional diffusion equation. J. Comput. Phys. **281**, 825–843 (2015)
14. W. McLean, K. Mustapha, Time-stepping error bounds for fractional diffusion problems with non-smooth initial data. J. Comput. Phys. **293**, 201–217 (2015)
15. Z. Wang, S. Vong, Compact difference schemes for the modified anomalous fractional sub-diffusion equation and the fractional diffusion-wave equation. J. Comput. Phys. **277**, 1–15 (2014)

Chapter 6
Requirements of MATLAB/Simulink for Signals

6.1 Development Tools for Modelling Special Processor Structures Based on MATLAB and Simulink

Owing to such qualities of MATLAB and Simulink, as an integrated development of algorithms and large numbers of data analysis functions, evaluation of an application for digital processing of signals (DPS) is considerably simplified and accelerated. Offering the engineers, a working intercourse language, MATLAB reduces the gap between an idea, scientific research and the final product. This open architecture allows working in interaction with other software products and systems in real time. Using MATLAB, the designer can check his ideas, assess tolerance limits and generate solutions that meet the strictest demands [1].

Matlab accelerates designing of applications owing to integration in a uniform environment such different plan such as working language with matrices, visual modelling, automatic generation of program code and additional software packages for many fields of knowledge. Engineers consider MATLAB an ideal instrument for signal processing. Its powerful language of matrix computations is natural for representation of signals and evaluation of algorithms for DPS. Writing of a program in MATLAB takes a small amount of time in comparison with programming on C/C++, without any loss in flexibility or quality.

Applied software packages of Matlab (Toolboxes) and Simulink is endless sources of ready to use functions, basis blocks for construction of models and visual instruments of working with signals [2, 3]. This ensures a wonderful base for user's own algorithms and products. Simulink, as a component of the MATLAB product package for signal processing, allows fast designing, modelling and testing of DSP systems using interactive visual modelling with the help of diagrams. Simulink helps analyse the work of algorithms already at the earliest stages of software designing. As the software designer gets close to final implementation of his plans, it does not get any complicated to modify his algorithm, for particularisation of application or further approximation to real conditions does not complicate the task of the software engi-

D. Singh et al., *Signal Processing Applications Using Multidimensional Polynomial Splines*, SpringerBriefs in Applied Sciences and Technology, https://doi.org/10.1007/978-981-13-2239-6_6

neer. MATLAB toolboxes contain newest algorithms, supplied with documentation and detailed user guides [4]. They help the designer be aware of the latest new products in digital signal processing field, such as wavelets or modern spectral analysis, and use such in his own researches (Fig. 6.1).

MATLAB and Simulink allow performing automated generation of software code of your applications. The Real-Time Workshop is capable of generating ANSI C standard codes for working with built-in patches. Matia Compiler converts the MATLAB language-evaluated algorithms into C/C++ coding, which in interaction with math libraries enables the creation of autonomous applications.

The teams of software evaluation engineers can interact using MATLAB as the language of intercourse and basic designing. This open system allows easy modification of the initial code, link outside software and data, separate ideas and software for working on PC, UNIX and Macintosh platforms.

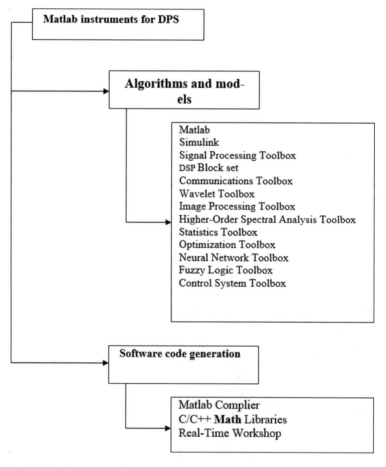

Fig. 6.1 MATLAB instruments for digital processing of signals

MATLAB and Simulink offer an alternative method to the traditional methods of designing applications for digital processing of signals. Whether you are adapting standard algorithms or creating your own, integrated into MATLAB/Simulink algorithms, MATLAB/Simulink accelerates the process of creating the application only because it enables experimenting with different approaches to the solution of the task [5, 6].

MATLAB language is ideal for programming of algorithms for digital processing of signals. Because the main element of the language is a massif, evaluation of algorithms in accuracy is such that you simply take the record of mathematical formulae. In addition, you can select algorithms from the wide range of checked collection of mathematical functions and MATLAB signal processing functions. As opposed to programming in C or C++, it is not necessary to start the software from zero or think how to link complicated libraries. Your searches in an interactive mode would take you a fully functional, wonderfully tuned algorithm. The Simulink block diagrams reflect a hierarchic model structure that simplifies designing of DSP systems and modelling of their behaviour. A Simulink block may represent a separate element of the system, a large subsystem or something in the middle of the two. Each block can be modified, tuned in accordance with the necessity and obtain the exact behaviour of the system for each specific case without going back to the traditional software designing. Using Simulink, one can interactively or pragmatically change the model parameters in the process of modelling [7]. The plug-in units are to be displayed on the screen for analysis of frequencies, etc., that enables analysing the work of the system on the go. Simulink can work with both: continuous and discrete systems. Therefore, it is possible to easily model complex systems that include time-varying components (subsystems) alongside with analogue components. There is also a possibility of including outside C or C++ or FORTRAN programs in user's own design, because dynamically linked object modules can be called from any MATLAB function or Simulink block. Use of such an approach can help store user's own checked development and create libraries for target applications.

The Simulink program is an application to the MATLAB package. In modelling with the use of Simulink, the visual programming principle is utilised, according to which, on the screen, the user creates a model of the installation and performs calculations from the standard blocks from the library. As opposed to the classical modelling methods, the user does not have to study the programming language in detail and the numeric mathematical methods; some general knowledge of working on the computer is sufficient, and, naturally, some knowledge of the subject field, where the user is currently working, Simulink is a sufficiently independent MATLAB instrument, and at working with Simulink, it is not absolutely necessary to know MATLAB and the rest of application. On the other hand, access to MATLAB functions and other instruments is left open and they can be used in Simulink. Part of the components of the packages has instruments that are built in Simulink (for instance, LTI Viewer of Control System Toolbox application—of the management system evaluation package). There are also additional block libraries for different usage fields (for instance, Power System Blockset—modelling of electrical devices, Digital Signal Processing Blockset—set of evaluation of digital devices, etc.). In working with Simulink, the user has an opportunity of modernising library blocks, creating his own libraries, and compiles new block libraries. In modelling, the user may select the method of differential equations and the method of model time (with a fixed or variable step). During modelling, there is a possibility to monitor the processes, active within the system. This is possible owing to the special monitoring installation that is part of Simulink library. Modelling results can be represented as curves or tables. The advantage of Simulink is also in the possibility of augmenting the block libraries with the help of subprograms, written in both MATLAB and C++, FORTRAN and Ada languages.

Simulink ensures extremely wide opportunities for creating signal processing programs for modern scientific and technical applications [8].

The powerful imitational real-time modelling subsystem that is connected to Simulink (subject to availability of additional hardware as computer expansion patches), represented by Real-Time Windows Target and Workshop expansion packages is a powerful device for management of real objects and systems [9]. The advantage of such modelling is its mathematical and physical visualisation (demonstration) level. In Simulink, not only model components can instruct fixed parameters, but they also include mathematical rations that describe the model behaviour.

Two types of signals can be generated in Simulink: continuous and discrete. For modelling of the continuous systems work with continuous base **time-based** type is recommended, while for modelling the work of discrete systems the **sample-shaped** type is recommended.

If a **time-based** model is installed, then the **sample time** can take the values:

- 0 (by default)—the block works in continuous mode.
- 0—the block works in discrete mode.
- −1—the block takes over the same mode as the accepting block.

Signal discretisation in continuous mode can be implemented using the **Zero-Order Hold** block.

The **Zero-Order Hold** block can be described s "a digitiser", i.e. part of ADC (analog-to-digital converter), responsible for discretisation of the signal. Sometimes, the Zero-**Order Hold** is called ADC. However, no quantisation is performed in **Zero-Order Hold** blocks.

The readout massifs of time moments and corresponding values of a signal can be exported from Simulink environment to the MATLAB environment using workspace block.

For the purposes of generation of given signals, we shall take the necessary blocks from the Simulink blocks library and compile a diagram, shown in Fig. 6.2.

By double-clicking on the oscilloscope block, activate the window that imitates the oscilloscope screen and initiates the model (**Start simulation** button). As a result, we obtain the sinusoid section image (Fig. 6.3).

Now, generate a section of a discrete harmonic signal in Simulink with the same parameters, used in MATLAB: amplitude 1, frequency 100 Hz, digitising frequency 5000 Hz, initial phase $\pi/2$ and the number of readouts 20 (Fig. 6.4).

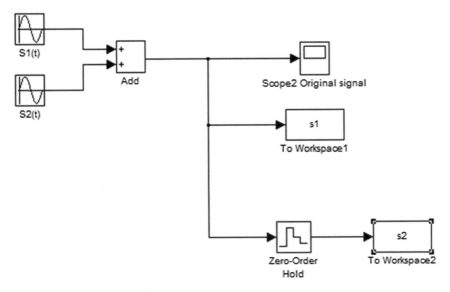

Fig. 6.2 Signal generation diagram in Simulink

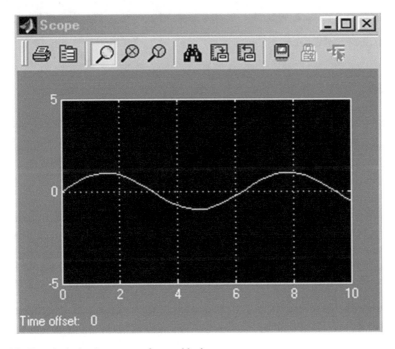

Fig. 6.3 Signal, obtained at output of scope block

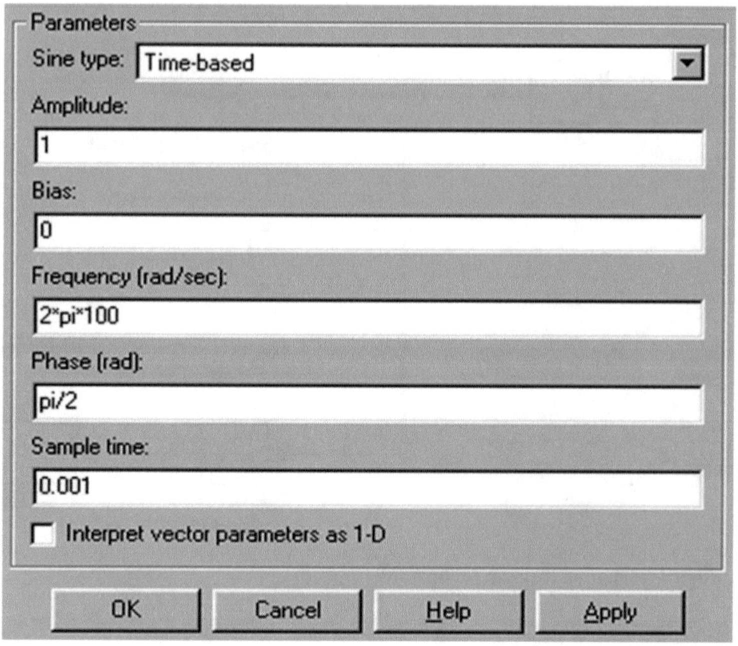

Fig. 6.4 Simulation parameters setup

Upon initiation of the model, the image appears on the oscilloscope window (Fig. 6.5).

Thus, Figs. 6.3 and 6.5 demonstrate the results of signal generation in Simulink.

6.2   Summary

The studies of the spline function methods indicate that as in the algorithmic computation of coefficients as well as in algorithms of recovery of signals there are frequent operations like parallel additions, multiplication and multiplication with accumulation. These operations are typical for digital processing of signals. Therefore, the algorithms, obtained because of application of the spline function methods, are convenient for the implementation of use in digital signal processors.

An analysis of existing hardware that intended for digital processing of signals indicated that Harvard's architecture and availability of hardware implemented special multiplication commands, parallel multiplication with accumulation offer an opportunity of wide use of modern digital signal processors for implementation of spline-recovery methods.

The MATLAB instrumental tools accelerate development of application owing to the integration of such different plan tools like the language for working with

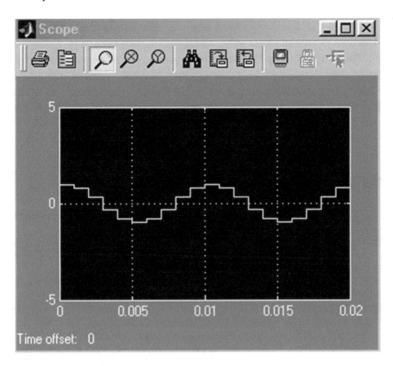

Fig. 6.5 Image on oscilloscope screen

matrixes, visual modelling, automatic generation of the software code and additional packages for various knowledge areas into one environment. Its powerful language of matrix computations is natural for representation of signals and development of algorithms for digital processing of signals. Additional packages of applied MATLAB (toolboxes) software and Simulink blocks are the richest sources of ready functions, basic blocks for construction of models and visual tools for work with signals. This ensures a wonderful base for user's own algorithms and software. As an integral part of MATLAB products package for signal processing, Simulink allows fast development, modelling and testing of digital processing of signals using interactive visual modelling with the help of diagrams. Simulink helps analyse the work of algorithms at the earliest stages of software development.

Some instrumental tools of modelling of the processes of recovery of one-dimensional and multidimensional signals for digital signal processors of Blackfin family of Analog Devices have been proposed. The instrument allows the designers to develop and perform debugging of applications. This environment includes an easy-to-use assembler. A C/C++ compiler and their executable functions libraries are includes mathematic and digital signal processing functions. A key feature of software development tools is the efficiency of the code that is written in C/C++ languages.

A class of digital signal processors has been analysed, and the Analog Devices ADSP Blackfin processors were selected for the implementation of spline-recovery methods. This processor family best fits for processing of the chosen class of signals. The cost of such processors are relatively low; and have parallel multipliers with accumulation and dynamic management of power source, performance rate and the required volume of memory storage capacity for processing of signals and supports integrated VisualDSP++ development environment.

References

1. National Instruments White Paper, Signal generation in Labview (2016). http://www.ni.com/white-paper/4087/en/
2. M. Dehghan, M. Abbaszadeh, A. Mohebbib, Error estimate for the numerical solution of fractional reaction-subdiffusion process based on a meshless method. J. Comput. Appl. Math. **280**, 14–36 (2015)
3. R. Ferré, Discrete convolution with modulo operations. Appl. Math. Lett. **4**(5), 13–17, ISSN 0893-9659 (1991). https://doi.org/10.1016/0893-9659(91)90135-I
4. M. Sławomir, Combination of the meshless finite difference approach with the Monte Carlo random walk technique for solution of elliptic problems. Comput. Math. Appl. **76**(4), 854–876, ISSN 0898-1221 (2018). https://doi.org/10.1016/j.camwa.2018.05.025
5. R. Sarmiento, F. Tobajas, V. de Armas, R. Esper-Chaín, J.F. López, J.A. Montiel-Nelson, A. Núñez, A Cordic processor for FFT computation and its implementation using gallium arsenide technology. IEEE Trans. Very Large Scale Integr. VLSI Syst. **6**(1), 18–30 (1998). http://www.iuma.ulpgc.es/~lopez/journals/IEEE%20TVLSI%2098.pdf
6. R.B. Abdessalem, S. Nejati, L.C. Briand, T. Stifter, Testing advanced driver assistance systems using multi-objective search and neural networks, in *Proceedings of the 31st IEEE/ACM International Conference on Automated Software Engineering (ASE 2016)* (USA, 2016), pp. 63–74. https://doi.org/10.1145/2970276.2970311
7. J.Y. Cao, Z.Q. Wang, A new numerical scheme for the space fractional diffusion equation, in *Frontiers of Manufacturing Science and Measuring Technology IV*. Applied Mechanics and Materials, vol. 599 (Trans Tech Publications, 2014), pp. 1305–1308. https://doi.org/10.4028/www.scientific.net/AMM.599-601.1305
8. X. Li, H. Li, B. Wu, A new numerical method for variable order fractional functional differential equations. Appl. Math. Lett. **68**, 80–86, ISSN 0893-9659 (2017). https://doi.org/10.1016/j.aml.2017.01.001
9. H. Liao, P. Lyu, S. Vong, Y. Zhao, Stability of fully discrete schemes with interpolation-type fractional formulas for distributed-order subdiffusion equations. Numer. Algorithms **75**(4), 845–878 (2016). https://doi.org/10.1007/s11075-016-0223-7

Chapter 7
Geophysical Application for Splines

7.1 Software for Processing of Geophysical Signals

During many geophysical studies, the scientist's efforts are oriented to search for reliable indicators of mineral deposits and seismic dangers. The indicators are abrupt changes, bursts/spikes or anomalies in one or other parameter, which may be used for prognosticating—forecasting a deposit and the quantity of minerals, and the locations, forces and time of future seismic events. The indicatives may be anomalous changes of electromagnetic and gravity fields, anomalous disturbances in the ionosphere, seismic noises, various acoustic vibrations, etc. [1].

Dozens of new methods have lately been proposed for prognosticating minerals and seismic events [2]. The results, obtained from these methods, are absolutely required for understanding the physical processes that lead to the seismic event and those that accompany such event, for the construction of physical and mathematical models of the interrelation of ongoing processes that are necessary for the practical implementation of prognosticating principles [3].

The functions of one argument, that are mathematical analogues of producible signals in geophysical studies and bench tests, can generally be represented as entire rational functions—multiples, fractional rational functions and transcendental functions—demonstrative logarithmic and trigonometric functions [4].

Therefore, the modelling was performed for elementary functions: harmonic, exponential and logarithmic functions, also for those functions that consist of their combinations. Figure 7.1 is the graphical representation of the results of approximation of function $y = \sin(\pi x)$ at interval $[-1.5; 1.5]$ by parabolic basic splines on the three-points formula.

The signals, obtained from magnetic and gravity sounding studies (for the purpose of prognosticating mineral deposits) and the signals obtained from seismic exploration works were used as real experimental data [5].

© The Author(s), under exclusive license to Springer Nature Singapore Pte Ltd. 2019   55
D. Singh et al., *Signal Processing Applications Using Multidimensional
Polynomial Splines*, SpringerBriefs in Applied Sciences and Technology,
https://doi.org/10.1007/978-981-13-2239-6_7

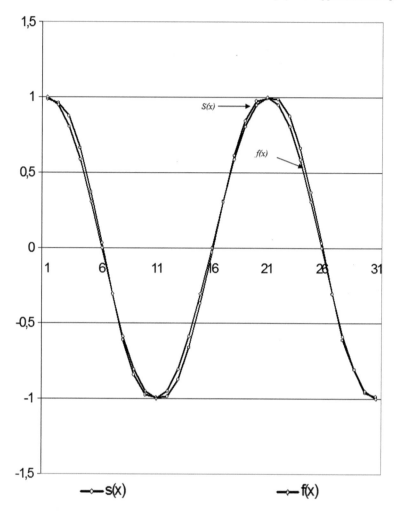

Fig. 7.1 Results of approximation of function $y = \sin(\pi x)$ at interval $[-1.5; 1.5]$ by parabolic basic splines on a three-points formula

In the first case, the frequencies are determined by the periods of signal selections, and particularly in magnetic sounding, it falls within the range of parts from a second to several seconds.

In the second case, the length of the signal is 10–400 mcs and frequency 10–1700 kHz. Usually, the frequency range of an indicator signal is divided into low frequency—from 10 to 50 kHz and high frequency—from 50 to 1700 kHz.

Figure 7.1 presents a record of geophysical signals, obtained from aerial-magnetic sounding studies. The following are shown on the coordinate axis:

Local time (LT) and world time (UT), geomagnetic latitude λ and parameter L show the position of the on-board registration devices over the source of the seismic event. The induction magnetic meter data were used for search of electric–magnetic disturbance over the seismic source (sensitivity $\approx 10^{-8}$ $(nT)^2/Hz$). The device registered a signal in the frequency range of 10 Hz–1 kHz. The magnetic meter signal was registered on seven channels with central frequencies of 10, 22, 47, 100, 216, 550 and 1000 Hz. The recording was performed at the time-day interval of to and after an earthquake. From the six fly-over attempts, three cases saw some disturbance of the electromagnetic radiation [6, 7]. Figure 7.2 illustrates the highest approximation of the on-board registrar with the epicentre, where the electromagnetic disturbances were observed at frequencies higher than 100 Hz.

When an effect is in place at a weak magnetic disturbance and at low latitudes, the probability of occurrence of natural radiations of magnetosphere origin is very low (i.e. noises). Therefore, for further work, we shall select a signal, presented in Fig. 7.2. The results of processing of this signal with parabolic splines are graphically presented in Fig. 7.2.

7.2 Generation of Electromagnetic and Acoustic Emissions

A mountainous massif is an electromechanical object. Its instability drops more under the influence of extreme dynamic loads, caused by a mass explosion. An impulse-nature large pressure is applied on the massif, which determines the specific feature of the development of redistribution of tensions [8]. The shock wave of the explosion changes the elastic, electric and other properties, which leads to the generation of electromagnetic (EMI—electromagnetic impulses) and acoustic (AE—acoustic emissions) signals in some areas of the massif [9]. This is associated with redistribution of the mechanical tension and electrical charges in the mountain rock. Due to this, the study of the electromagnetic and acoustic activity of the massif is real for the purpose of prognosticating of rock bumps [10].

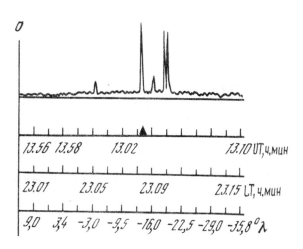

Fig. 7.2 Record of a geophysical signal, obtained during aero-magnetic sounding studies

During experiments for evaluation of shock danger of the mountain massif after a massive explosion, reception of electromagnetic signals is performed by a broadband magnetic antenna (20 kHz–2 MHz) of the device for the radio-impulse method of exploration, while for acoustic signals by five 60 kHz piezoceramic transducers. For registration of electromagnetic and acoustic signals, a package of radio-impulse method devices is used. This method ensures synchronous registration of electromagnetic and acoustic signals. Simultaneous recording of EMI analogue signals is taken during such experiments.

The experiment must be conducted in conditions of complete exclusion of all electromagnetic and acoustic noises. Therefore, there is actually no interference of any noises during studies of electromagnetic and acoustic activity of the mountain massif under study, or they do not influence the reliability of the obtained information.

During processing, the energetic parameters of EMI and AE signals were calculated, and the curves of their time behaviour were drawn. After a massive explosion, several shocks occur, which are compared with the time values of anomalous changes of electromagnetic and acoustic emissions [11]. Most informative parameters, which are anomalously changed before the class four dynamic event, occurred in 2.2 h after the mass explosion—ratio of full energies of EMI and AE signals is calculated for each accumulation time period (Fig. 7.3). In this case, an advancing effect of the anomalous rise of electromagnetic activities is observed in comparison with the anomalous increase of the acoustic activity prior to the mountainous/rock shock.

During the experiment, most of the EMI-registered signals are high-frequency signals (over 300 kHz) in lengths no greater than 10 mcs. In addition, some low-frequency EMI signals are registered (10–50 kHz) at lengths of 100–400 mcs.

At the initial stage of preparation of the rock shock, some multiple high-frequency EMI signals are registered as packages of from 300 to 1000 mcs lengths, i.e. the number of high-frequency signals in each package varied from 50 to 200. The signal spectres are of similar form, while the lengths are about 4–6 mcs. A spectral maximum

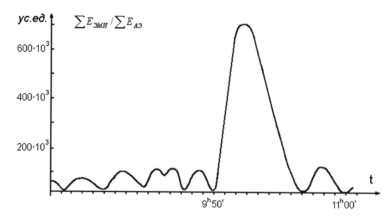

Fig. 7.3 Variations of energy parameters of EMI and AE

is noted at a frequency of about 1.6–1.7 MHz. As the rock shock preparation is developed, the spectres of registered high-frequency signals are complicated. Some high-frequency signals with several spectral maximums occur.

It characterised that during the process of preparation of the rock shock, the high-frequency signals were registered most of the time, mainly as packages and single low-frequency signal, and a few minutes prior to the rock shock, some individual high-frequency and low-frequency signals were registered, and some small packages of high-frequency EMI signals were also registered. Prior to the shock, the frequency of spectral maximums of high-frequency signals is relocated to the area of lower frequencies 1.6–1.8 MHz—500–800 kHz. The similar parameter of low-frequency signals does not get subjected to special changes. The number of high-frequency EMI signals drops prior to the shock, while the number of low-frequency signals increases. We introduced a new PRC parameter, equal to the EMI HF–LF signals ratio. The value of this parameter increases anomalously a few dozen minutes before the rock shock and then started dropping. The high-frequency signal amplitude prior to the shock gradually increases from 300 to 800 mv, while the amplitude of low-frequency signals does not change at all, though some increase of value dispersion is noted. For 4–5 min prior to the rock shock, the length of high-frequency signals increases (from 4–6 to 18–26 mcs), while the length of low-frequency signals does not notably change.

Figure 7.3 illustrates the results of processing of the geophysical signals using the spline-function methods, obtained as the result of magnetic sounding exploration. The average quadratic deviation of recovery of the geophysical signal is 0.582.

Figure 7.4 illustrates the geophysical signal f(x) and the recovered geophysical signal S(x).

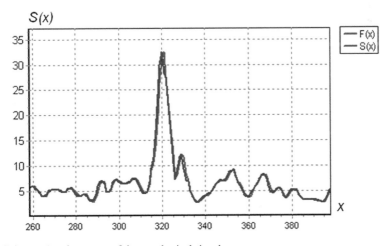

Fig. 7.4 Results of recovery of the geophysical signal

The following conclusions can be drawn based on the completed experiment and processing of real geophysical signals, obtained by way of experimental observations [12]:

(1) Some non-synchronous increase of energy generation of EMI and AE is noted prior to the dynamic event;

(2) The effect of advancing of period of high electromagnetic activity is confirmed in comparison with the period of increase of the acoustic activity prior to the dynamic event;

(3) It was established that some increase in the number of low-frequency EMI signals occurs prior to the rock shock, amplitude of the high-frequency EMI signals and their length, and drop of the frequency of the high-frequency EMI signals;

(4) A prognostication parameter of EMI–PRC was introduced that takes anomalous values a few dozen minutes before the rock shock.

(5) A spline method was developed for the determination of anomalies of seismic signal–indicator of seismic events. This method is based upon the use of a mathematical device for cubic splines.

In scientific studies' practice, there is a frequent task whereby it is necessary to recovery the general nature of an event or a process using experimental data.

7.3   Splines' Implementation of Geophysical Studies Data

The classical solution of this task is a selection of a function from the acceptable multiplicity, which best approximates the combination of experimental data [13]. The average quadratic error value is often used for assessment of the degree of the quality of approximation of the function to experimental data. In such cases, a practical implementation of such approach is the smallest quadrates method. But the application of the smallest quadrates method leads to the solution of the algebraic equation systems. For those systems that function in real-time scales, including for multi-layer seismic tests, development of new efficient methods deems to be necessary [14]. One of the ways of solution of such issue is the use of the spline methods of experimental data approximation.

The broad use of spline methods in analysis and processing tasks for seismic and geophysical signals are explained by their being a universal instrument of approximation and, in comparison with other mathematical methods at equal informational and hardware costs, they ensure higher accuracy levels [15].

Figure 7.5 shows the results of approximation of data from geophysical studies with parabolic B-splines on a three-points formula. Figure 7.6 illustrates the results of smoothing of experimental data from geophysical studies with parabolic B-splines using the smallest quadrates at $l = 128$, $N = 44$ (average quadratic error is 0.014, relative error 0.070).

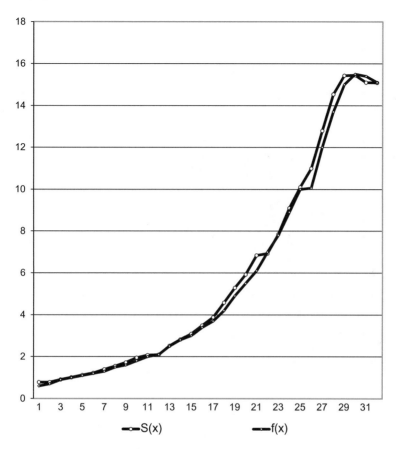

Fig. 7.5 Results of approximation of geophysical studies data using parabolic B-splines on the three-points formula

7.4 Summary

This chapter delineates some applications of splines stemming from their use in signal processing and approximation. Algorithms and software were developed for studying approximation using basic splines, which enabled the determination of the parameters of interpolation splines and the computation of B-spline coefficients on local formulae. They are also helpful in the recovery of functions and signal values and in their smoothening using the basic spline on smallest quadrates method. Some software and algorithms were also written for studying basic spline approximation and transformation algorithms on DSP. A Simulink model was made for the parallel computation structure based on cubic basic splines. Based on the proposed spline-function methods, some software tools for the processing of signal and experimental data were developed.

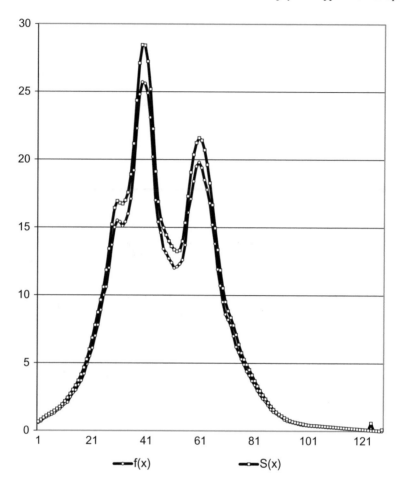

Fig. 7.6 Results of smoothing of geophysical studies data with parabolic B-splines using the smallest quadrates method

They were developed on the VisualDSP++ simulator, for the Blackfin-BF533 digital signal processors for Analog Devices. These tools are easily implementable of DSP and may be used to process other types of signals. Mathematical tools and cubic splines were used to determine anomalies of complex signal indicators of seismic events. This use of spines will lead to significant improvement in the accuracy of results and a reduction in the computation costs. Another application of the spline method of analysis are processing and determination of anomalies using spectral basic signals in raid structures to access the fatigue, stability and strength of railway rails.

References

1. R. Silva-Ortigoza, C. Márquez-Sánchez, M. Marcelino-Aranda, M. Marciano-Melchor, G. Silva-Ortigoza, R. Bautista-Quintero, E.R. Ramos-Silvestre, J.C. Rivera-Díaz, D. Muñoz-Carrillo, Sci. World J. **2013**(723645), 17 (2013). http://dx.doi.org/10.1155/2013/723645

2. J. Liu, H. Guo, Y.-L. Jiang, Y. Wang, High order numerical algorithms based on biquadratic spline collocation for two-dimensional parabolic partial differential equations. Int. J. Comput. Math. (2018). https://doi.org/10.1080/00207160.2018.1437260

3. D. Singh, H. Zaynidinov, H.-J. Lee, Piecewise-quadratic Hartmuth basis functions and their application to problems in digital signal processing, Special Issue: Next Generation Networks (NGNs). Int. J. Commun. Syst. **23**(6–7), 751–762. (2010). https://doi.org/10.1002/dac.1093

4. F. Liao, L. Zhang, S. Wang, Numerical analysis of cubic orthogonal spline collocation methods for the coupled Schrödinger-Boussinesq equations. Appl. Numer. Math. **119**, 194–212 (2017)

5. H.P. Quach, T.C.P. Chui, Low temperature magnetic properties of Metglas 2714A and its potential use as core material for EMI filters. Cryogenics **44**(6–8), 445–449, ISSN 0011-2275 (2004). https://doi.org/10.1016/j.cryogenics.2004.01.006

6. G. Yi, J. Wang, K.-M. Tsang, X. Wei, B. Deng, C. Han, Spike-frequency adaption of a two-compartment neuron modulated by extracellular electric fields. Biol. Cybern. **109**(3), 287–306 (2015). https://doi.org/10.1007/s00422-014-0642-2

7. H. Paasche et al.: MuSaWa: multi-scale S-wave tomography for exploration and risk assessment of development sites. Adv. Technol. Earth Sci. 95–114 (2014). https://doi.org/10.1007/978-3-319-04205-3_6

8. J. Tang, H. Cheng, L. Liu, Using nonlinear programming to correct leakage and estimate mass change from GRACE observation and its application to Antartica. J. Geophys. Res.: Solid Earth banner **117**(B11), 148–227 (2012). https://doi.org/10.1029/2012JB009480

9. E. Quirós, Á.M. Felicísimo, A. Cuartero, Testing Multivariate Adaptive Regression Splines (MARS) as a method of land cover classification of Terra-Aster satellite images. Sensors **9**(11), 9011–9028 (2009). https://doi.org/10.3390/s91109011

10. A. Carrassi, S. Vannitsem, State and parameter estimation with the extended Kalman filter: an alternative formulation of the model error dynamics. Q. J. R. Meteorol. Soc. **137**(655), 435–451 (2011). https://doi.org/10.1002/qj.762

11. M.M. Khader, A new fractional Chebyshev FDM: an application for solving the fractional differential equations generated by optimisation problem. Int. J. Syst. Sci. **46**(14), 2598–2606 (2015). https://doi.org/10.1080/00207721.2013.874508

12. A. Eicker, T. Mayer-Gürr, K.-H. Ilk, E. Kurtenbach, in *Regionally Refined Gravity Field Models from In-Situ Satellite Data*. System Earth via Geodetic-Geophysical Space Techniques (Springer, 2010), pp. 255–264. https://doi.org/10.1007/978-3-642-10228-8_20

13. X. Yang, H. Zhang, D. Xu, WSGD-OSC scheme for two-dimensional distributed order fractional reaction-diffusion equation. J. Sci. Comput. **76**(3), 1502–1520 (2018). https://doi.org/10.1007/s10915-018-0672-3

14. H. Zhang, X. Yang, The BDF orthogonal spline collocation method for the two-dimensional evolution equation with memory. Int. J. Comput. Math. **95**(10), 2011–2025 (2018). https://doi.org/10.1080/00207160.2017.1347259

15. H. Chen, C. Zhang, Convergence and stability of extended block boundary value methods for Volterra delay integro-differential equations. Appl. Numer. Math. **62**(2), 141–154, ISSN 0168-9274 (2012). https://doi.org/10.1016/j.apnum.2011.11.001

Chapter 8
Spline Evaluation for Railways

8.1 Analysis of Spline Implementation in Railway Rails

A specific feature of the modern stage of development of experimental studies is their automation based on the broad implementation of computer tools [1]. Automation of experimental studies and development and implementation of new methods of data processing allow expanding the possibilities of experimental devices and organise a new direction of studies, improving the labour production rate of researchers and efficiency of the use of expensive equipment [2–4].

Bench tests of new materials and complex objects subject to the impact of climatic, mechanical, acoustic and other factors are an important part of experimental test work. Such bench test work is conducted in the construction of railways, in machinery, aviation and space equipment production, electronic equipment, device production and at a number of other industries [5]. The main purpose of bench tests is improving the quality, reliability, stability and equipment longevity at a decrease of their material consumption and cost by way of producing the conditions, equal to the conditions, met in real operational conditions.

The extensive operation of railway rails showed that currently there are no sufficient scientific and experimental data for the evaluation of rail stability.

Use of this method is explained by the necessity in:

- Comparative assessment of workability of rails in sales/tender processes, when the purpose is a selection of the most optimal option from the economic point of view out of several rail options, proposed for purchase;
- Assessment of operating stability of rails that are already installed on the railway.

The introduction of this methodology must ensure justified and reliable selection of rails at purchases of a new batch of rails abroad and expedient operation of existing rails.

Based on the study and summary of information on the studied rails in laboratory conditions and polygon tests of rails at the experimental ring and on railway net-

© The Author(s), under exclusive license to Springer Nature Singapore Pte Ltd. 2019
D. Singh et al., *Signal Processing Applications Using Multidimensional Polynomial Splines*, SpringerBriefs in Applied Sciences and Technology,
https://doi.org/10.1007/978-981-13-2239-6_8

work, the methodology of calculation of relative wear resistance, resistance against contact fatigue damages and chipping/crumbling of different rails [6] (carbonaceous, alloyed, oxidised and modified rails) is subjected to different types of thermal and thermo-mechanical treatment. The energy consumption rate of the material at plastic deformation is used as the criterion for assessment of operability of rails.

In modern sophisticated times, the operating requirements, applied to the rails, are determined by several parameters. An optimal combination of these parameters creates the highest operability rate of rail steel. Often, the structural strength and wear resistance of rail used to be associated with its yield strength and hardness. However, rails that had similar strength specifications, but the resistance of different rails with different microstructure types differed on their strength [7]. Correlation between the static strength of railheads and their operability exists within each type of microstructure.

This work proposes the methodology of assessment of relative operability of rail steel types, subjected to different types of thermal treatment, which have a different structure and mechanical properties [8]. A relative value of concealed energy of deformations is accumulated in deforming amounts of the material at the destruction point, i.e. the energy capacity of the material.

The necessity for further improvement of efficiency of control systems at the simultaneous reduction of cost and number of operators is obvious. Improvement of used technologies of railway diagnostics assumes development of computer methods of analysis of control data, of identification of defective sections and assessment of their degree of danger. Consequently, the directions of signal processing, registration and demonstration methods necessary for development are gaining more importance by the day [9].

It was established that the main sources of unexpected abrupt fatigue damages of rails, wheel pair treads, locomotive power parts and other goods are the availability of zones of concentration of internal tensions of the metal (TC zones), caused by manufacturing technologies. The rail and wheel pair tread manufacture plants do not have any efficient methods and means of control over technological defects of manufacturing and residual tensions [10].

The traditional methods and means of operating control (magnetic and ultrasound fault detectors) enable to identify existing developed defects. These control tools do not ensure diagnostics of rails and wheel pairs at the predestruction stage and, therefore, cannot guarantee the safety of transportation on railways. The magnetic wagon fault detectors are based on reading magnetic field of dispersion that is formed in the zone of a developed defect at artificial magnetisation of the rail with a constant magnetic field [11].

The spectral theory of B-splines was used for processing of the signal, obtained in the process of studies.

The idea of finite basics of the B-splines are in signal processing tasks in the analysis of such function as Kotelnikov–Shannon's expansion nucleus as two variables [12]:

$$K(\omega, t) = \frac{\sin(\omega t)}{\omega t} \tag{8.1}$$

The local nature of B-splines and finiteness of sections, at which the continuous signals are usually determined in real validity, are the basis for moving to the solution of the discretion task from signal models with a finite spectre to spectral representation of signals, which are of the class of integer functions of frequency argument [13]. The unique feature of polynomial B-splines is that their amplitude spectral characteristics correspond to the distinct and sufficiently simple analytical description, which has a lot common with the description of Kotelnikov–Shannon's nucleus row:

$$F_{B0}(\omega) = h\, B_m(0) \left(\frac{\sin(\omega h/2)}{\omega h/2} \right)^{m+2} \tag{8.2}$$

where h—discretisation step; $B_m(0)$—amplitude (maximal value) of B-spline of level m.

It has been noted above that the random continuous function $f(x)$ given in the final section $x \in [a, b]$ and to be approximated on this section by the polynomial spline $S_m(x)$ may be represented as a linear sum:

$$f(x) \cong S_m(x) = \sum_{i=-1}^{n+1} b_i B_{m,i}(x), \tag{8.3}$$

where i—current number of the spline node ($i = -1, 0, 1, \ldots, n + 1$); b_i—succession coefficients of B-splines that approximate the function.

Representation of amplitude spectral density of the succession, which consists of the final number of rectangular impulses (zero-level B-splines) and operator expression of succession, which includes the final number of pair intersecting triangular impulses (the first-level B-splines) is easily obtained in the analytical form [14, 15]. By replacing the complex variable p to $j\omega$, the operations can move to the spectral plane from the operator expression. If this approach is expanded for applying to the final succession of non-intersecting B-splines of random integer level m with varying amplitudes, the amplitude spectral density of such succession may be described by the following formula:

$$F_{\Sigma B}(\omega) = F_{B0}(\omega) \left| \sum_{i=0}^{n} b_i \exp(-j i\, \omega(m + 1)h) \right|. \tag{8.4}$$

It contains complex exponential multipliers that take into consideration the effects of delays of each following B-splines to each previous spline by the step h, which is also the distance between two adjacent spline nodes:

$$h = x_{i+1} - x_i = \text{const} \quad (i = 0, 1, 2, \ldots, n).$$

The B-splines should not intersect each other; i.e. they must be orthogonal. This requirement is automatically satisfied for zero-level basic splines, rectangular impulses, because each of such has its own separate carrier. In other words, the spectral densities of the first-level B-splines may be computed separately for groups of integer and odd impulses. Inside each of these groups, the impulses are orthogonal [16]. The second-level B-splines can also be separated into orthogonal element groups, if the zero elements are combined with the third and the first element is combined with the fourth element. For the purposes of satisfying the orthogonality requirements, the cubic B-splines are grouped on principle "zero" element to be combined with the fourth, the first element to be combined with the fifth element, etc.

8.2 Stability Evaluation of Railways

The completed work showed unique opportunities of the new diagnostics method: without any special magnetisation, using two to four gauges, without any direct contact with the surface to be tested, perform express analysis of rail condition and determine the areas, predisposed to damages. In addition, scanning can be done at the train motion speed [17].

An abrupt local change of the magnetic field Hp in Fig. 8.1 corresponds to the zone of a developing defect (fissure on the railhead). It should be noted that the tension concentration line ($Hp = 0$) indicates the direction of development of the identified fissure.

Therefore, a spline method of analysis, processing and determination of anomalies of rail structure changes is proposed based on the use of spectral features of the polynomial B-spline. This method allows assessing the fatigue and prognosticating the stability and strength of railway rails [18].

8.3 Discussion

Some software tools of signal and experimental data processing were developed based on the proposed spline-function methods on VisualDSP++ simulator for Blackfin ADSP-BF533 digital signal processors from Analog Devices. These software tools are easily implemented on DSP and can be used for processing of other types of signals. A spline method of determination of anomalies of complex seismic signal indicators of seismic events was developed based on the use of the mathematical tool of cubic splines. The use of spline functions leads to the improvement of accuracy of the results and to significant reduction of computation costs. A spline method of analysis, processing and determination of anomalies of rail structures, based on the use of spectral basic signals, was developed. The developed method allows assessing the fatigue and prognosticating the stability and strength of railway rails.

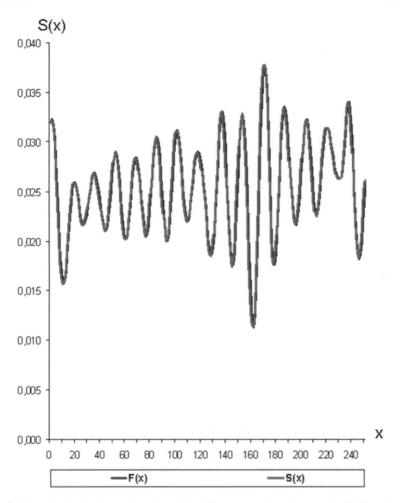

Fig. 8.1 Results of signal approximation obtained from measuring the magnetic field in a known rail zone

References

1. S. Bosse, M. Koerdt, D. Schmidt, Robust and adaptive signal segmentation for structural monitoring using autonomous agents, in *Proceddings the 4th International Electronic Conference on Sensor and Applications*, vol. 2, issue 3 (2017) p. 105. https://doi.org/10.3390/ecsa-4-04917
2. S.M. de Lima, L.V. Vareda, J.B.L. Liborio, High performance concrete applied to storage system buildings at low temperatures. Mater. Res. **11**(2), 121–130 (2008). http://www.scielo. br/pdf/mr/v11n2/a03v11n2.pdf
3. M. Koudstaal, F. Yao, From multiple Gaussian sequences to functional data and beyond: a Stein estimation approach. J. R. Stat. Soc., B **80**, 319–342 (2018). https://doi.org/10.1111/rssb. 12255

4. B. Shillingford et al., Large scale visual speech recognition, arXiv:1807.05162V2 []cs.CV]
    (2018). https://arxiv.org/pdf/1807.05162.pdf
5. R. Sharma, L. Vignolo, G. Schlotthauer, M.A. Colominas, H.L. Rufiner, S.R.M. Prasanna,
    Empirical mode decomposition for adaptive AM-FM analysis of speech: a review. Speech
    Commun. **88**, 39–64, ISSN 0167-6393, (2017). https://doi.org/10.1016/j.specom.2016.12.004
6. P.H.M. Bovendeerd, P.A. Van Steenhoven, F.N. van de Vosse, G. Vossers, Steady entry
    flow in curved pipe flow. J. Fluid Mech. **177**, 233–246 (1987). https://doi.org/10.1017/
    s0022112087000934
7. A. Sotiras, C. Davatzikos, N. Paragios, Deformable Medical image registration: a survey. IEEE
    Trans. Med. Imaging **32**(7), 1153–1190 (2013). http://doi.org/10.1109/TMI.2013.2265603.
    (*PMC*. Web. 28 Sept. 2018)
8. D.B. Keele, Jr. (Don), Log sampling in time and frequency: preliminary theory and appli-
    cation. Audio Engineering Society Convention (1994). http://www.aes.org/e-lib/browse.cfm?
    elib=6297
9. D. Schillinger, E. Rank, An unfitted hp-adaptive finite element method based on hierarchical
    B-splines for interface problems of complex geometry. Comput. Methods Appl. Mech. Eng.
    200(47–48), 3358–3380, ISSN 0045-7825, (2011). https://doi.org/10.1016/j.cma.2011.08.002
10. K.E. Emblem et al., Vessel architectural imaging identifies cancer patient responders to anti-
    angiogenic therapy. Nat. Med. **19**(9), 1178–1183 (2013). http://doi.org/10.1038/nm.3289
11. N. Gonga-Saholiariliva, Y. Gunnell, C. Petit, C. Mering, Techniques for quantifying the
    accuracy of gridded elevation models and for mapping uncertainty in digital terrain anal-
    ysis. Prog. Phys. Geogr.: Earth Environ. **35**(6), 739–764 (2011). https://doi.org/10.1177/
    0309133311409086
12. F. Gensun, Whittaker–Kotelnikov–Shannon Sampling theorem and aliasing error. J. Approx.
    Theory **85**(2), 115–131, ISSN 0021-9045, (1996). https://doi.org/10.1006/jath.1996.0033
13. A.R. Amiri-Simkooei, M. Hosseini-Asl, A. Safari, Least squares 2D bi-cubic spline approxi-
    mation: theory and applications. Measurement **127**, 366–378, ISSN 0263-2241, (2018). https://
    doi.org/10.1016/j.measurement.2018.06.005
14. H.N. Zaynidinov, M.B. Zaynutdinova, E.S. Nazirova, Methods of reconstucting signals
    based on multivariate spline. Eur. J. Comput. Sci. Inf. Technol. **3**(2), 56–63 (2015).
    http://www.eajournals.org/wp-content/uploads/Methods-of-reconstructing-signals-based-on-
    multivariate-spline2.pdf
15. C.H. Garcia-Capulin, F. Cuevas, G. Trejo-Caballero, H. Rostro, A hierarchical genetic algo-
    rithm approach for curve fitting with B-splines. Genet. Program. Evolvable Mach. **16**, 151–166
    (2014). https://doi.org/10.1007/s10710-014-9231-3
16. H. Zhou, T. Wenyi, W. Deng, Quasi-compact finite difference schemes for space fractional
    diffusion equations. J. Sci. Comput. **56**(1), 45–66 (2013). http://dx.doi.org/10.1007/s10915-
    012-9661-0
17. H. Zhang, X. Yang, The BDF orthogonal spline collocation method for the two-dimensional
    evolution equation with memory. Int. J. Comput. Math. **95**(10), 2011–2025 (2018). https://doi.
    org/10.1080/00207160.2017.1347259
18. G. Zhang, A. Xiao, Exact and numerical stability analysis of reaction-diffusion equations with
    distributed delays. Front. Math. China **11**(1), 189–205 (2016). https://doi.org/10.1007/s11464-
    015-0506-7

Printed in the United States
By Bookmasters